T0320974

Games, Puzzles,
and Computation

Games, Puzzles, and Computation

Robert A. Hearn
Erik D. Demaine

A K Peters, Ltd.
Wellesley, Massachusetts

Editorial, Sales, and Customer Service Office

A K Peters, Ltd.
888 Worcester Street, Suite 230
Wellesley, MA 02482
www.akpeters.com

Copyright © 2009 by A K Peters, Ltd.

All rights reserved. No part of the material protected by this copyright notice may be reproduced or utilized in any form, electronic or mechanical, including photocopying, recording, or by any information storage and retrieval system, without written permission from the copyright owner.

Library of Congress Cataloging-in-Publication Data

Hearn, Robert A.
  Games, puzzles, and computation / Robert A. Hearn, Erik D. Demaine.
     p. cm.
  Includes bibliographical references and index.
  ISBN 978-1-56881-322-6 (alk. paper)
  1. Problem solving—Mathematical models. 2. Games—Mathematical models. 3. Logic, Symbolic and mathematical. I. Demaine, Erik D., 1981– II. Title.

  QA63.H35 2009
  510–dc22

                                                                    2009002069

Cover images: See Figures 1.2(a), 10.11(b), and C.12(a).

Printed in India
13 12 11 10 09   10 9 8 7 6 5 4 3 2 1

Acknowledgments

I would like to thank a few of the very many people who contributed directly or indirectly to my part in the making of this book: Michael Albert, Cyril Banderier, Eric Baum, Jake Beal, Elwyn Berlekamp, John Conway, Martin Demaine, Gary Flake, Aviezri Fraenkel, Greg Frederickson, Ed Fredkin, Martin Gardner, Shafi Goldwasser, J. P. Grossman, Richard Guy, Charles Hearn, Lerma Hearn, Michael Hoffmann, Michael Kleber, Tom Knight, Charles Leiserson, Norm Margolus, Albert Meyer, Marvin Minsky, Chet Murthy, Richard Nowakowsi, Ed Pegg, Ivars Peterson, Tom Rodgers, Aaron Seigel, Michael Sipser, Gerald Jay Sussman, John Tromp, Patrick Winston, David Wolfe, and Warren Wood.

This book arose out of my thesis work at MIT. Thus, it would not have been possible without Erik Demaine, who first interested me in tackling the complexity of sliding-block puzzles, and through whose mentoring and collaboration that initial result led to a stream of related results and eventually this book.

Special thanks are due the staff of A K Peters, most particularly Charlotte Henderson, who displayed amazing patience as deadlines slipped and who offered many valuable suggestions and improvements.

My deepest thanks go to my wife Liz, who is the reason I was at MIT in the first place. Finally, Liz, I've made something of all that fooling around with games and puzzles!

—R.A.H.

I would like to thank, at a broader level, the people who influenced the whole body of research. Most people studying the mathematics of games and puzzles, and the two of us in particular, were heavily influenced by Martin Gardner. His 25 years of *Scientific American* articles and dozens of books showed the world how these fields could be combined. His influence continues today through the Gathering for Gardner meetings, organized by Tom Rodgers, giving a meeting place for many enthusiasts of games and puzzles and of mathematics and computer science. Other key meeting places have been provided by the combinatorial games community, in particular Elwyn Berlekamp, Richard Nowakowski, and David Wolfe. For me this began with the Second Combinatorial Games Theory Workshop and Conference in 2000, whose proceedings led to the book *More Games of No Chance*. Next came the Dagstuhl Seminar on Algorithmic Combinatorial Game Theory in 2002, which I helped organize, that specifically brought together people who work on algorithms and people who work on combinatorial games. These early meetings played an important role in getting this research area off the ground.

It has been exciting to go on this particular adventure with Bob Hearn. We started working together on the complexity of games in 2001 when I arrived at MIT, and our collaboration has been productive. Bob has been excitedly pushing the frontiers of the interplay between games, puzzles, and computation ever since we discovered Nondeterministic Constraint Logic, and I am happy that the culminated research is now embodied as both his PhD thesis and this book.

Finally, I would like to thank my father, Martin Demaine, whose passion for life and learning in general, and for games and puzzles in particular, ultimately brought me here. We have been sharing and collaborating throughout my life, all the way to this research, and beyond.

—E.D.D.

Contents

Introduction

This book is about games people play and puzzles people solve, viewed from the perspective of computer science—in particular computational complexity. Over the years, we have found increasingly deep connections between games, puzzles, and computation. These connections are interesting to us from multiple perspectives. As game players and puzzle solvers, we find underlying mathematical reasons that games and puzzles are challenging, which perhaps explain why they are so much fun. As computer scientists, we find that games and puzzles serve as powerful models of computation, quite different from the usual models of automata and circuits, offering a new way of thinking about computation.

This book has three main parts, and different parts may be of interest to different readers.

Part I (Games in General) describes a framework we have developed for studying the connections between games, puzzles, and computation, called *constraint logic*. This framework defines one simple prototypical game/puzzle that can be interpreted in a variety of settings. We can vary the number of players: one-player puzzles, two-player games, multiplayer team games, or, at the other extreme, zero-player automata. We can also vary how many moves for which the game lasts, or whether the players can hide information (like cards) from each other. In each such category of games, we prove that the corresponding form of constraint logic is the computationally most difficult game in that category, making it a natural point of reference from the computer-science perspective. This part of the book is fairly technical, building a mathematical foundation for particular constraint logics and establishing their computational complexity. Readers

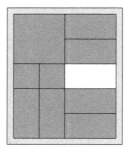

Figure 1.1. Dad's Puzzle, one of the earliest (c. 1909) and most popular sliding block puzzles [71]. The solver must slide the nine rectangular pieces within the 4×5 box to get the large square into the lower-left corner. The shortest solution takes a whopping 59 moves.

uninterested in the details, however, can simply read the summaries in the two short opening chapters, 2 and 3.

Part II (Games in Particular) applies the constraint-logic framework to real games and puzzles that people play. The approach is to take a real game or puzzle and show that it is computationally as hard as the corresponding form of constraint logic, making the real game/puzzle also computationally most difficult in its category. The intuition is that most "interesting" games are the most difficult in their class, so as a result we end up with many "equally difficult" games (when held up to the fairly course grain of computational complexity theory). What is interesting is that many real games and puzzles can be closely modeled within the constraint-logic framework, making it fairly easy to establish these complexity results.

Constraint logic started out as a tool for understanding the complexity of sliding-block puzzles, such as the puzzle shown in Figure 1.1. Our pursuit was motivated by a problem posed by Martin Gardner [71]: "These puzzles are very much in want of a theory. Short of trial and error, no one knows how to determine if a given state is obtainable from another given state...." The first application of the constraint-logic framework, which we will see in Section 9.3, shows that these puzzles have no such general theory, in a computational sense: no efficient procedure can tell whether a given state is obtainable from another, assuming standard beliefs in computational complexity. From there, the theory of constraint logic grew to increasing generality, capturing more and more types of real games and culminating in this book.

The third main part of this book, Appendix A (Survey of Games and Their Complexities), serves as a reference guide for readers interested in the computational complexity of particular games, or interested in open problems about such complexities. While Part II establishes the complexity of many games, it focuses on applications of the constraint-logic framework, and currently not all game-complexity results fit this framework. Appendix A surveys all known results, in addition to highlighting many open problems.

The rest of this introduction gives the reader some basic background on the two main concepts of this book—games/puzzles and complexity—followed by a more detailed overview of the constraint-logic framework.

1.1   What is a Game?

The term *game* means different things to different people in different fields. Our use intends to capture the kinds of games that people play, including board games like Chess, Checkers, and Go; card games like Poker and Bridge; one-player puzzles like Rush Hour, Peg Solitaire, and Sliding Blocks; and zero-player automata like John Conway's Game of Life.

Common to all of these games are four main features: positions, players, moves, and goals. Every game we consider has finitely many possible *positions*: board configurations, card distributions, piece arrangements, etc. In computer-science terminology, our games have a *bounded state*, a finite amount of information that defines the current situation. Some number of *players* manipulate the game position by individual *moves*. The players take turns in some order; the next player to move can be viewed as part of the game position. During each turn, the current player has a clear list of allowable moves (defined by the rules of the game) and picks one of them. The move transforms the game position into some other game position, in particular advancing to the next player in whatever order is determined by the game. Players may not be able to observe certain parts of the game position, allowing players to have *hidden states* such as cards in a hand, but this hidden state should not prevent a player from determining their allowable moves. Each player has a *goal*: to reach a game position with a particular property. The first player to reach their goal *wins*. We generally assume *optimal play*: players try to win as best they can given the available information. Although we do not directly consider games with randomness such as dice rolls in this book, we can model such phenomena by supposing that one player plays randomly instead of following optimal play (as in [131]).

This informal definition is related to several types of games studied in a variety of fields. To provide some context for our study of games, we summarize the related results and differences in these fields.

Combinatorial Game Theory. One closely aligned study of games is *combinatorial game theory*, as in the two classic books *Winning Ways* [8] and *On Numbers and Games* [27]; see also the more recent introduction *Lessons in Play* [3] and the research collections *Games of No Chance* I–III [128–130]. The bulk of this study considers two-player games of *perfect information*, where every player knows the entire state of the game and the moves available to each player—no hidden cards, random dice rolls, etc. Combinatorial game theory builds a beautiful theory of such perfect-information two-player games, revealing a rich mathematical structure. Perhaps most surprising is the connection to number systems: real numbers are special cases of Conway's "surreal numbers," which in turn are special cases of games, and basic addition carries over to the general case of games.

Perfect information has the attractive consequence that, in principle, a player could determine the optimal move to make by simulating the entire game execution, trying every possible move by each player (assuming the game is finite). Algorithmic combinatorial game theory aims to understand when there are better strategies than such brute force, and combinatorial game theory builds a useful collection of tools for understanding such optimal strategies in games. In many if not most interesting games, however, optimal game play is a difficult computational problem, and proving such results is our purpose in studying the complexity of games.

Economic Game Theory. A less related study of games is *(economic) game theory*, as pioneered by the work of John von Neumann [169] and John Nash [127]. Here, two or more selfish players participate in an economic event (game), often framed as a single round in which each player simultaneously chooses a strategy (or, often, a probability distribution of strategies), and the score (outcome) for each player is a given function of these strategies. In this context, there is generally no clear optimum strategy, either globally or for each player. There is, however, a clear set of optimal strategies for one player when given the strategies of other players, and if all players simultaneously follow such a strategy, the strategies are in *Nash equilibrium*. Nash [127] proved that all games have such an equilibrium, with the idea that players' strategies will eventually converge to one. On the other hand, theoretical computer scientists [23, 34] recently established that finding a Nash equilibrium is computationally intractable (formally, PPAD-complete), so players of normal computational power will in general take a long time to converge to a Nash equilibrium. More generally, economic game theory studies a wide variety of different notions of equilibria and the properties they possess.

The games we consider are both more specialized and more general than what is traditionally addressed by game theory: more specialized because we are concerned only with determining the winner of a game, and not with other issues such as maximizing payoff, cooperative strategies, etc.; more general because game theory is concerned only with the interactions of two or more players, whereas we will consider games with only one player (puzzles) and even with no players at all (simulations).

Computational Complexity Theory. Computational complexity theory studies the general notion of computation, which in its broadest sense can encompass all of the games we consider. Indeed, the perspective of this book is that games serve as powerful models of computational devices. Standard complexity theory, however, focuses on one kind of computation, the *Turing machine* [166]. From our perspective, the standard Turing machine can be viewed as a zero-player game or automaton, following simple rules to change the position (state), ending with either a winning (accepting) or losing (rejecting) state. The two critical resources of such a machine/game are *time*—the number of moves that can be made before the game ends— and *space*—the amount of information that can be remembered through the game position (pieces on the game board, etc.). Impressively, with respect to both of these resources, the Turing machine is essentially equivalent to (or more powerful than) all known physical computational devices, built and conceived, with the exception of quantum computers, which seem to offer some additional computational power [149]. Nonetheless, further insight into the structure of computational problems can be obtained from more powerful models of computation, and much such insight has already come from the idea of adding players to form a game.

Adding a single player that can choose which moves to make, aiming to arrive at a winning position, corresponds to the classic idea of *nondeterministic computation* [166]. If such a one-player puzzle is time limited—that is, the player can make only a reasonable (polynomial) number of moves before the puzzle ends—then we obtain the complexity class NP, immortalized in the famous P vs. NP problem. Restated in game-theoretic terms, the P vs. NP problem asks whether an optimal puzzle player can be simulated efficiently by a zero-player automaton. The complexity community broadly believes that the answer is "no": computing the winning strategy to a puzzle takes a long (exponential) time in general. This belief is the foundation for real-life puzzles such as Eternity and Eternity II [35, 163]—whose solution is worth millions in prize money—and is part of the basis for modern cryptography upon which everyday banking relies. P vs. NP is also one of the Clay Mathematics Institute's seven Millennium Prize Problems [105], whose solution wins $1,000,000.

Adding two players that compete to reach a winning position corresponds to the less famous but also classic idea of *alternating computation* [22]. Alternation refers to the opposite roles of the two players: for the first player to win, there must be *some* move with the property that *all* moves by the second player have *some* move for the first player with the property that *all* moves by the second player .... Although more powerful than nondeterminism, alternation is fairly well-understood: alternation for a reasonable (at least polynomial) amount of time corresponds to computing within roughly the same (up to polynomial factors) amount of space [22]. As a result, we will often see computational space requirements arising in the complexity of games.

Several other models of computation have been introduced, in particular to capture aspects of games that people play (unlike the models above that were introduced out of independent interest). *Privacy* [136, 137] adds the possibility that the players cannot see parts of the game position, as with hidden cards. *Solitaire automata* [111] consider the effect of nondeterminism and privacy in one-player puzzles when combined with the idea of an arbitrary (shuffled) initial position. *Team games* with private information [133] lead to the possibility of infinite computation using only finitely many perceivable resources, a topic we will explore in Chapter 7.

Another direction, studied throughly by complexity theory but not in this book, is *randomness*. Examples include games against Nature [131], interactive proof systems [79], Arthur-Merlin proof systems [5], stochastic automata [55], and probabilistic game automata [24]. See [26] for an overview of such models.

1.2   Computational Complexity Classes

Our general goal is to analyze the computational complexity of games and puzzles. More precisely, for each game, we define one or more computational problems or questions about the game whose answer is a single bit: "yes" or "no." Most typically, we ask whether a particular player wins under optimal play from a given game position, i.e., whether the player can (force a) win. Each such problem belongs to certain complexity classes, and our goal is to find the most specific class into which a particular problem fits. We give now an informal description of the complexity classes of interest to us; for formal definitions, refer to Appendix B.

In general, these complexity classes study how the necessary resources—time and space—grow in terms of the *input size*. For problems about games, the input is the game description—pieces, cards, board, etc.—whose size we measure as binary bits of information. We can think of an *algorithm* (Turing machine) as a zero-player automaton defined by simple deterministic rules that transform the game position from one to the next, eventually producing an answer. As described above, the *time* is the number of such moves before the automaton answers, and the *space* is the amount of information stored by the automaton (its size).

Time. We start by characterizing in terms of time. The class P consists of all problems solvable in *polynomial time*, that is, all problems solved by some algorithm in time that is at most linear, quadratic, cubic, or similar in the input size. If n represents the input size, a general polynomial might look like $5n^4 + 3n^2 + 10n - 1$. As long as the problem can be solved within such time, for some polynomial, then the problem is in P. Problems in P are generally considered easy to solve, because polynomials scale well. A simple example of a game in P is the game of Nim, where it is easy to determine which player will win under optimal play.

Similarly, the class $EXPTIME$ consists of all problems solvable in *exponential time*: 2^n, 5^n, 2^{n^2}, or in general $2^{p(n)}$ where $p(n)$ is some polynomial. This class is much larger, including most two-player games such as Chess, and it is usually easy to solve a problem in exponential time. Note that EXPTIME contains easier problems too, in particular, all of P.

Space. On the other hand, we can characterize by space. As mentioned above, these classes are equivalent to characterizing by time but for alternating (two-player) computation, so they frequently arise in two-player games. The class $PSPACE$ consists of all problems solvable in polynomial space. This class is the analog of P but measuring space instead of time. Similarly, $EXPSPACE$ consists of all problems solvable in exponential space.

We can relate the space classes with the time classes as follows. An optimal algorithm never uses more space than time. Thus, for example, every problem in P is also in PSPACE. Also, any (deterministic) algorithm that uses s space can never use more than exponential-in-s time without repeating a position, which would cause an infinite loop and never correctly answer the problem. Thus, for example, every problem in PSPACE is also in EXPTIME.

Nondeterminism. Next, we consider allowing nondeterminism in the algorithm, i.e., allowing the automaton to be driven by a single player making choices among possible rules to apply. Here, the answer produced by the algorithm becomes ambiguous, because it depends on what choices the player makes. To make the answer well defined, we dictate that the overall output is "yes" whenever the player has *some* sequence of choices that produces a "yes" answer. Equivalently, we can think of the player as winning when it reaches a "yes" answer, and we can define the player to follow an optimal strategy. In this sense, a nondeterministic algorithm can be thought of as extremely lucky: whenever it needs to make a decision, it by definition makes the correct choice.

The class NP consists of all problems that can be solved in polynomial time by such a nondeterministic algorithm. Similarly, we can define *NPSPACE* for the nondeterministic analog of PSPACE, and *NEXP-TIME* for EXPTIME. One nice result is that PSPACE and NPSPACE are identical: nondeterminism does not help when measuring space and not time [146]. Nondeterministic computation is at least as powerful as regular deterministic computation, so, for example, every problem in P is also in NP. On the other hand, nondeterministic computation can be simulated by trying both choices of each decision in turn, which takes exponentially more time, but about the same amount of space. Thus, for example, every problem in NP is also in PSPACE.

To summarize, we can write $X \subseteq Y$ to denote that every problem in X is also in Y (X is contained in Y) and conclude that

$$P \subseteq NP \subseteq PSPACE \subseteq EXPTIME \subseteq NEXPTIME \subseteq EXPSPACE.$$

Each of these classes represents computation at a certain level of ability: given certain time or space resources, and possibly given superpowers such as nondeterminism. There are also some problems that cannot be solved in general by *any* algorithm in finite time. We call these problems *undecidable*, and they effectively represent computation without bounded resources. Even more surprising is that, in a certain technical sense, *most* problems are undecidable, even though it is rare to encounter such problems in practice. Indeed, we will see that such problems arise from team games.

Completeness. It is usually easy to show that a game (or more precisely, a computational problem/question about a game) falls into a particular complexity class: just exhibit a simple algorithm to do so. But how do we know when we are done, that we cannot push the game into any lower complexity class? Computational complexity provides a powerful tool for side-stepping this issue. For each complexity class X, we call a problem X-*hard* if it is about as hard as every problem in X. (Here, we ignore polynomial factors in the difficulty.) We call a problem X-*complete* if it is both X-hard and in X. Thus, for example, NP-complete problems are among the hardest problems in NP, so they must not be in any strictly easier complexity class. Whether P = NP is of course a major open problem, but assuming they are even slightly different, NP-complete problems are not in P.

In general, for each game problem we consider, our goal is to find a class X for which the problem is X-complete. We then know we are done, having eliminated the problem from any strictly smaller complexity class—whatever those turn out to be, pending solutions to outstanding conjectures in complexity theory. (All classes we have described are conjectured to be different, except the already-mentioned property that nondeterminism does not affect space classes. All we know so far is that P and EXPTIME are different.)

1.3 Constraint Logic

The central theme of this book is constraint logic. Constraint logic is one simple generic game that can be easily adapted to all different types of games: games with zero players, one player, two players, or two teams of players; with polynomially bounded length or unbounded length; with and without hidden information. The constraint-logic framework enables a uniform treatment of computational ability arising from these many game types. It can be seen as a game-oriented analog of Turing machines from complexity theory, which have served as a universal model of computation that can be easily restricted to model any complexity class in the time and space hierarchies described above. We believe that our alternative approach to understanding computation has several advantages, particularly when studying games that people play.

Simplicity through Graphs. Constraint logic is simple, being describable in a few sentences. The game board is any weighted undirected graph: a collection of *vertices* (represented by dots) connected by *edges* (represented by line segments or curves), where every vertex and edge is labeled with a *weight* of 1 or 2. A board position is an *orientation* of this graph—specifying a direction that each edge points—such that the total weight of edges directed into each vertex is at least that vertex's desire (weight). A move is the reversal of an edge orientation that results in a valid board position. A computation is a sequence of moves. Generally the goal is to reverse a particular distinguished edge.

This model differs in several ways from traditional logical models of computation such as circuit or formula satisfiability. Constraint logic is simple, being modeled directly around graphs. Circuits and formulae can be represented as graphs with sufficient augmentation (e.g., distinguishing variables from clauses/gates). But it is difficult to see the actual computation taking place in such a graph, whereas the dynamics of constraint logic defines the entire state in the orientation of the graph. We find this purely combinatorial view, without any explicit logical values, much easier to work with in many cases.

Completeness. Each type of game has a natural complexity class into which it falls. What is interesting is that, for each game type that we have studied, its constraint logic is among the hardest problems in its complexity class. For example, both one-player unbounded games and two-player bounded games are essentially always in PSPACE, and the corresponding versions of constraint logic are PSPACE-complete. A more surprising result is that, for team games of imperfect information, constraint logic is undecidable: no algorithm can play the game perfectly. This game is the first undecidable game that has a finite number of positions, behaves deterministically, and has players alternate turns. See Section 7.2 for further discussion about this result.

Part I proves many such completeness results, forming the foundation for the theory of constraint logic. This foundation makes constraint logic a suitable starting point for proving that other games are hard, a topic to which we now turn.

Applications to Real Games. It is well known that many one-player puzzles and two-player games are complete within their natural complexity class. For example, Peg Solitaire is a bounded-length puzzle and NP-complete [168]; Sokoban and Rush Hour are unbounded-length puzzles and PSPACE-complete [33,48,56,85]; Hex and Othello are bounded-length two-player games and PSPACE-complete [106,140]; and generalized Chess, Checkers, and Go are unbounded-length two-player games and EXPTIME-complete [58,143,145]. For many more examples, refer to the survey in Appendix A. Most of these proofs are complicated and specific to the game being analyzed, reducing from appropriate forms of formula satisfiability.

Our primary motivation for developing the constraint-logic model of computation is that it is much closer in flavor to many existing games, making it easier to prove the complexity of a game by relating to its corresponding type of constraint logic. We have successfully developed such reductions for many different games: bounded-length one-player TipOver and Hitori (Sections 9.1 and 9.2); unbounded-length one-player Sliding Blocks, the Warehouseman's Problem, plank puzzles, Sokoban, Rush Hour, triangular Rush Hour, Push-2-F, Sliding Tokens, and hinged-dissection reconfiguration (Sections 9.3–9.11); and bounded-length two-player Amazons, Konane (Hawaiian Checkers), and Cross Purposes (Chapter 10). In the cases of Sokoban, Rush Hour, and the Warehouseman's Problem, which were proved PSPACE-complete previously, our proofs based on constraint logic are simpler and (except for Rush Hour) establish hardness for more specific forms of the games. The complexities of many of the other games were open problems for several years; for example, the complexity of sliding blocks was posed by Martin Gardner 40 years ago [71].

The constraint-logic framework provides a host of tools for making it easy to prove the completeness of a game within a particular complexity class. Such a proof only needs to show how to implement two kinds of constraint-graph vertices, one representing a kind of AND computation and another representing a kind of OR computation. (In fact, even simpler forms

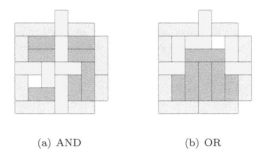

(a) AND   (b) OR

Figure 1.2. Constraint logic gadgets showing PSPACE-completeness of sliding-block puzzles.

of these vertices suffice to establish completeness.) The proof also needs to be able to construct edges that connect two vertices. Furthermore, in most cases, the constraint-logic framework guarantees that the graph is planar, so a game-specific proof does not need to worry about edges crossing. In contrast, crossover gadgets are the most complicated part of most previous completeness proofs for two-dimensional games; with constraint logic, they come for free so are unnecessary to build.

As an illustrative example of the power of the constraint-logic framework, Figure 1.2 shows the entire construction required to prove that sliding-block puzzles are PSPACE-complete. Thus, the solution to this 40-year-old problem becomes an almost-trivial "proof by two pictures." Readers interested in many more such reductions (in addition to this one) are referred to Part II.

Outside of games, others have used the unbounded-length one-player form of constraint logic to establish the complexity of airport planning [96], steel slab stacking [109], finding paths between graph colorings [14], and morphing parallel graph drawings [155].

Recently we discovered an application of zero-player constraint logic to evolutionary graph theory [45].

1.4   What's Next?

The next two chapters, 2 and 3, detail constraint logic and describe the specific games it defines. They provide the necessary background both for studying and for using constraint logic, so we recommend them to all readers. The rest of Part I is then for readers interested in the theory of constraint logic and why the games serve as representative complete problems in their respective classes. Next, Part II is for readers interested in the application of that theory to real games and puzzles that people play. Finally, Appendix A is for those interested in finding out the complexity status of their favorite game or puzzle, or for those interested in open problems.

Games in General

In Part I of this book we develop the constraint-logic model of computation in its various flavors. These comprise instances in a two-dimensional space of game categories, shown in Figure I.1. The first dimension ranges across zero-player games (deterministic simulations), one-player games (puzzles), two-player games, and team games with private information. The second dimension is whether the game has (polynomially) bounded length. In all cases, the games use bounded space; the basic idea is that a game involves pieces moving around, being placed, or being captured, on a board or other space of fixed size.

	Zero player (simulation)	One player (puzzle)	Two player	Team, imperfect information
Unbounded	PSPACE	PSPACE	EXPTIME	RE (undecidable)
Bounded	P	NP	PSPACE	NEXPTIME

Figure I.1. Table of constraint-logic categories and complexities. Each game type is complete for the indicated class. (After [133].)

Chapter 2 defines the general constraint-logic model of computation. Chapter 3 defines all the various flavors of constraint logic and describes their complexities. Chapters 4–7 develop notions of game ranging from deterministic simulations to team games of private information, and provide corresponding complexity results for appropriate versions of constraint logic.

Chapter 8 explores some of the implications of the results in Part I.

The Constraint-Logic Formalism

The general model of games we will develop is based on the idea of a *constraint graph*; by adding rules defining legal moves on such graphs we get *constraint logic*. In later chapters the graphs and the rules will be specialized to produce games with different numbers of players: zero, one, two, etc. A game played on a constraint graph is a computation of a sort, and simultaneously serves as a useful problem to reduce to other games to show their hardness.

In the game complexity literature, the standard problem used to show games hard is some kind of game played with a Boolean formula. The Satisfiability problem (SAT), for example, can be interpreted as a puzzle: the player must existentially make a series of variable selections, so that the formula is true. The corresponding model of computation is nondeterminism, and the natural complexity class is NP. Adding alternating existential and universal quantifiers creates the Quantified Boolean Formulas problem (QBF), which has a natural interpretation as a two-player game [158, 159]. The corresponding model of computation is alternation, and the natural complexity class is PSPACE. Allowing the players to continue to switch the variable states indefinitely creates a formula game of unbounded length, raising the complexity to EXPTIME, and so on. Most game hardness results (e.g., Instant Insanity [142], Hex [53, 140], Generalized Geography [147], Chess [58], Checkers [145], Go [114, 143]) are direct reductions from such formula games or their simple variants, or else even more explicit reductions directly from the appropriate type of Turing machine (e.g., Sokoban [33]).

One problem with such reductions is that the geometric constraints typically found in board games do not naturally correspond to any properties of the formula games. By contrast, the constraint-logic games we will present are all (with one exception) games played on planar graphs, so that there is a natural correspondence with typical board game topology. Furthermore, the required constraints often correspond very directly to existing constraints in many actual games. As a result, the various flavors of constraint logic are often much more amenable to reductions to actual games than are the underlying formula games. As evidence of this, we present a large number of new games reductions in Part II. The prototypical example is sliding-block puzzles, where the physical constraints of the blocks—that two blocks cannot occupy the same space at the same time—are used to implement the appropriate graph constraints.

Constraint logic also seems to have an advantage in conceptual economy over formula games. Formula games require the concepts of variables and formulas, but constraint-logic games require the single concept of a constraint graph. In the reductions we will give from formula games of various types to equivalent constraint logics, the variables and the formulas are represented uniformly as graph elements. This conceptual economy translates to simpler game reductions; often fewer gadgets must be built to show a given game hard using constraint logic.

Appendix B reviews Boolean formulas and the Satisfiability and Quantified Boolean Formulas problems; other formula games are defined in the text as they are needed.

2.1 Constraint Graphs

A *constraint graph* is an oriented graph, with edge weights $\in \{1, 2\}$. An edge is then called *red* or *blue*, respectively. The *inflow* at each vertex is the sum of the weights on inward-directed edges. Each vertex has a nonnegative *minimum inflow*. A legal configuration of a constraint graph has an inflow of at least the minimum inflow at each vertex; these are the *constraints*. A legal move on a constraint graph is the reversal of a single edge's orientation, resulting in a legal configuration. Generally, in any game, the goal will be to reverse a given edge, by executing a sequence of moves. In multiplayer games, each edge is controlled by an individual player, and each player has his own goal edge. In deterministic games, a unique sequence of reversals is forced. For the bounded games, each edge may only reverse once.

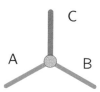 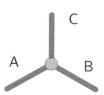

(a) AND vertex. Edge C may be directed outward if and only if edges A and B are both directed inward.

(b) OR vertex. Edge C may be directed outward if and only if either edge A or edge B is directed inward.

Figure 2.1. AND and OR vertices. Red (light gray) edges have weight 1, blue (dark gray) edges have weight 2, and vertices have a minimum inflow constraint of 2.

A constraint graph is an abstraction of the general notion of a game board, but it is also natural to view a game played on a constraint graph as a computation. Depending on the nature of the game, it can be a specific type of computation, e.g., a deterministic computation, or a nondeterministic computation, or an alternating computation. The constraint graph then *accepts* the computation just when the game can be won.

AND and OR Vertices. Certain vertex configurations in constraint graphs are of particular interest. A vertex with minimum inflow constraint 2 and incident edge weights of 1, 1, and 2 behaves as a logical AND, in the following sense: the weight-2 (blue) edge may be directed outward if and only if both weight-1 (red) edges are directed inward. Otherwise, the minimum inflow constraint of 2 would not be met. We will call such a vertex an *AND vertex*.

Similarly, a vertex with incident edge weights of 2, 2, and 2 behaves as a logical OR: a given edge may be directed outward if and only if at least one of the other two edges is directed inward. We will call such a vertex an *OR vertex*. AND and OR vertices are shown in Figure 2.1. Blue edges are drawn thicker than red ones as a mnemonic for their increased weight.

Directionality; FANOUT. As implied above, although it is natural to think of AND and OR vertices as having inputs and outputs, there is nothing enforcing this interpretation. A sequence of edge reversals could first direct both red edges into an AND vertex, and then direct its blue edge outward; in this case, we will sometimes say that its *inputs* have *activated*, enabling its *output* to activate. But the reverse sequence could equally well occur. In this case we could view the AND vertex as a splitter, or *FANOUT* gate: directing the blue edge inward allows both red edges to be directed outward, effectively splitting a signal.

In the case of OR vertices, again, we can speak of an active input enabling an output to activate. However, here the choice of input and output is entirely arbitrary, because OR vertices are symmetric.

Circuit Interpretation. With these AND, OR, and FANOUT vertex interpretations, it is natural to view a constraint graph made with these vertex types as a kind of digital logic network, or circuit. (See Figure 4.8 for examples of such graphs.) One can imagine signals flowing through the graph, as outputs activate when their input conditions are satisfied. This is the picture that motivates our description of constraint logic as a model of computation, rather than simply as a set of decision problems. Indeed, it is natural to expect that a finite assemblage of such logic gadgets could be used to build a sort of computer.

However, several differences between constraint graphs and ordinary digital logic circuits are noteworthy. First, digital logic circuits are deterministic. With the exception of zero-player constraint logic, a constraint-logic computation exhibits some degree of nondeterminism. Second, with the above AND and OR vertex interpretations, there is nothing to prohibit "wiring" a vertex's "output" (e.g., the blue edge of an AND vertex) to another "output," or an "input" to an "input." In digital logic circuitry, such connections would be illegal and meaningless, whereas they are essential in constraint logic. Finally, although we have AND- and OR-like devices, there is nothing like an inverter (or NOT gate) in constraint logic; inverters are essential in ordinary digital logic.

This last point deserves some elaboration. The logic that is manifested in constraint graphs is a *monotone* logic. By analogy with ordinary real functions, a Boolean formula is called monotone if it contains only literals, ANDs, and ORs, with no negations. The reason is that when a variable changes from **false** to **true**, the value of the formula can never change from **true** to **false**. Likewise, constraint logic is monotone, because inflow is a monotone function of incident edge orientations. Reversing an edge incident at a given vertex from in to out can never enable reversal of another edge at that vertex from in to out—that is what would be required by a NOT vertex. One of the more surprising results about constraint logic is that monotone logic is sufficient to produce computation, even in the deterministic case.

Flake and Baum [56] require the use of inverters in a similar computational context. They define gadgets ("both" and "either") that are essentially the same as our AND and OR vertices, but rather than use them as primitive logical elements, they use their gadgets to construct a kind of dual-rail logic. With this dual-rail logic, they can represent inverters, at a higher level of abstraction. We do not need inverters for our reductions, so we may omit this step.

AND and OR Subtypes; Basis Vertices. For some of the game categories, there can be many subtypes of AND and OR vertex, because each edge may have a distinguishing initial orientation (in the case of bounded games), and a distinct controlling player (when there is more than one player). These features break symmetries in generic vertex behavior. For example, in a bounded game, an AND vertex with the blue edge initially directed out and the red edges initially directed in can only function as a FANOUT, and not as a logical AND; once the red edges have reversed out, they can never reverse in again. Therefore, we call this AND subtype a FANOUT. However, we do not give a distinct subtype name to vertices where the red edges start directed out, and the blue edge in; these are still simply AND vertices. It should always be clear from context whether the initial edge orientation matters.

It turns out that for all the different kinds of constraint logic, it is sufficient to consider constraint graphs containing only AND and OR vertices. However, for some kinds of constraint logic it is useful to additionally employ a few other simple vertex types. One example is the CHOICE vertex, described below. For every kind of constraint logic, we give the best set of such "basis vertices" to work with. These are chosen so as to enable the easiest reductions from constraint logic to other games and puzzles.

If AND and OR vertices are always sufficient, why use other vertex types? Doesn't that simply create more gadgets that have to be implemented to perform a reduction? The answer is that sometimes many AND and OR subtypes are required if we wish to only use ANDs and ORs. Often we can require fewer or simpler gadgets by using other vertex types.

2.2   Planar Constraint Graphs

Crossover gadgets are a common requirement in complexity results for games and puzzles, and can be among the most difficult gadgets to design. (For example, the crossover gadget used in the proof that Sokoban is PSPACE-complete [33] is quite intricate.) It turns out that every constraint graph has an equivalent planar constraint graph,[1] which means that a reduction from a given game or puzzle to a constraint-logic game need not address how to cross edges. The specific form of the equivalent planar graph will vary depending on the type of constraint logic; a different planarity proof is given for each in the following chapters. But most of the proofs rely on the same underlying construction.

The fundamental crossover gadgets are shown in Figure 2.2(a) and Figure 2.2(b). In addition to AND and OR vertices, Figure 2.2(a) contains

---

[1]Bounded Deterministic Constraint Logic is the sole exception.

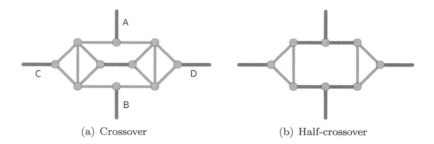

(a) Crossover   (b) Half-crossover

Figure 2.2. Planar crossover gadgets.

red-red-red-red vertices; these need any two edges to be directed inward to satisfy the inflow constraint of 2. The "half-crossover gadget" in Figure 2.2(b), which does use only ANDs and ORs, may be substituted in for each red-red-red-red vertex (using the red-blue conversion described in the following section).

The relevant property of the crossover gadget is that each of the edges A and B may face outward if and only if the other faces inward, and each of the edges C and D may face outward if and only if the other faces inward.

The proofs that specific kinds of constraint logic have planar equivalents use these gadgets, and additionally address issues such as edges being allowed to reverse only once, edges being controlled by distinct players, and so on.

2.3   Constraint-Graph Conversion Techniques

Various subgraph equivalences will be useful to us over and over, so we describe them here. Note that all of these equivalences strictly apply only under the generic constraint-logic rules; we must additionally observe the rules of each specific game type when applying them.

CHOICE Vertices. A *CHOICE* vertex, shown in Figure 2.3(a), is a vertex with three incident red edges and an inflow constraint of 2. The constraint is thus that at least two edges must be directed inward. If we view A as in input edge, then when the input is inactivated, i.e., A points down, then the outputs B and C are also inactivated, and must also point down. If A is then directed up, either B or C, but not both, may also be directed up. In the context of a game, a player would have a choice of which path to activate.

The subgraph shown in Figure 2.3(b) has the same constraints on its A, B, and C edges as the CHOICE vertex does. Suppose A points down. Then

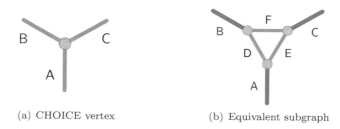

(a) CHOICE vertex   (b) Equivalent subgraph

Figure 2.3. CHOICE vertex conversion.

D and E must also point down, which forces B and C to point down. If A points up, D and E may as well (using vertex A-D-E as a FANOUT). F may then be directed either left or right, to enable either B or C, but not both, to point up.

The replacement subgraph still may not be substituted directly for a CHOICE vertex, however, because its terminal edges are blue, instead of red. This brings us to the next conversion technique.

Degree-2 Vertices. Viewing constraint graphs as circuits, we might want to connect the output of an OR, say, to an input of an AND. We cannot do this directly by joining the loose ends of the two edges, because one edge is blue and the other is red. However, we can get the desired effect by joining the edges at a red-blue vertex with an inflow constraint of 1. This allows each incident edge to point outward just when the other points inward—either edge is sufficient to satisfy the inflow constraint.

We would like to find a translation from such red-blue vertices to subgraphs using only ANDs and ORs. However, there is a problem: in ANDs and ORs, red edges always come in pairs. The solution is to provide a conversion from *two* red-blue vertices to an equivalent subgraph. The conversion is shown in Figure 2.4. Clearly, the orientations shown for the edges in the middle satisfy all the constraints except for the left and right vertices; for these, an inflow of 1 is supplied, and either the red or the blue edge is necessary and sufficient to satisfy the constraints. Note that the degree-2 vertices are drawn smaller than the AND/OR vertices, as an aid to remembering that their inflow constraint is 1 instead of 2.

What if there is a single leftover red-blue vertex that cannot be paired? This cannot happen in a graph using only ANDs, ORs, and red-blue vertices: a red edge incident at a red-blue vertex must be one end of a chain of red edges ending at another red-blue vertex. But it could happen if the graph uses CHOICE vertices. In that case we will supply the extra red edge (to turn a red-blue vertex into an AND) from a loose red edge (see below).

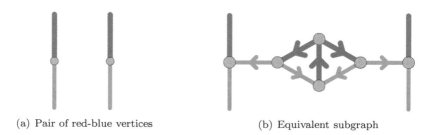

(a) Pair of red-blue vertices   (b) Equivalent subgraph

Figure 2.4. Red-blue vertex conversion. Red-blue vertices, which have an inflow constraint of 1 instead of 2, are drawn smaller than other vertices.

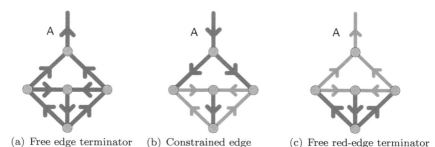

(a) Free edge terminator   (b) Constrained edge   (c) Free red-edge terminator
                              terminator

Figure 2.5. How to terminate loose edges.

These red-blue conversions are generally not strictly necessary: in most reductions to actual games, one builds AND- and OR-like gadgets that could easily be directly connected together anyway. That is, adding red-blue vertices to the basis set for a particular type of constraint logic generally does not make game reductions any more difficult. However, it is simpler and cleaner not to require them.

It will occasionally be useful to use blue-blue and red-red vertices, as well as red-blue. Again, these vertices have an inflow constraint of 1, which forces one edge to be directed in. A blue-blue vertex is easily implemented as an OR vertex with one loose edge that is constrained to always point away from the vertex (see below). Red-red edges will only occur in zero-player games. However, in that case special timing considerations also arise, so we will defer discussion of red-red vertices for now.

Loose Edges. Often only one end of an edge matters; the other need not be constrained. Properly, every edge in a graph must have a vertex at each end, but we can use "loose edges" in constraint graphs by assuming there is a vertex with an inflow constraint of 0 at the unattached end. We do not draw a vertex dot at all at that end, because the constraint is 0.

To translate such an edge into ANDs and ORs, the subgraph shown in Figure 2.5(a) suffices. If we assume that edge A is connected to some other vertex at the top, then the remainder of the figure serves to embed A in an AND/OR graph while not constraining it.

Similarly, sometimes an edge needs to have a permanently constrained orientation. The subgraph shown in Figure 2.5(b) forces A to point down; there is no legal orientation of the other edges that would allow it to point up.

If the graph is allowed to use CHOICE vertices, we can implement a free red edge as in Figure 2.5(c). Again, A is unconstrained by this subgraph.

Constraint-Logic Games

In this chapter we describe all the specific kinds of constraint-logic games in detail. In each case—from bounded zero-player games all the way to unbounded team games—we show how an appropriate kind of constraint logic can be viewed as a canonical game for that category. We state, without proving, the computational power (equivalently, computational-complexity class) of each game type, and give the simplest set of "basis vertex" types needed to show that specific games are hard. Armed with this information the reader may, if he desires, skip ahead to Part II fully prepared to understand the given reductions. All of the results stated here are summarized even more concisely in Table D.1.

The remainder of Part I of the book contains the formal proofs that the different kinds of constraint logic have the stated complexities. Strictly speaking, the rest of Part I and Part II both consist primarily of hardness reductions. (There is also some background and history on game complexity in the remainder of Part I.) The difference is that in Part I, the reductions are from several different kinds of problems of known complexity to constraint-logic games, and in Part II the reductions are from constraint-logic games to several different kinds of real-world games. But because the constraint-logic games are on one side or the other of the reduction in each case, the two cases have a different flavor. Part I may be viewed as "theory," and Part II as "application."

We'll begin at the beginning, with zero-player bounded games, the least interesting of the game types we consider. But the most basic or "pure" type of constraint logic is actually Nondeterministic Constraint Logic, corresponding to unbounded one-player games; all other kinds of constraint logic are natural specializations or generalizations of this basic problem. It is formally defined as follows (from Section 3.2.2):

NONDETERMINISTIC CONSTRAINT LOGIC (NCL)

Instance: Constraint graph G, edge e in G.

Question: Is there a sequence of moves on G that eventually reverses e?

This "boxed-problem" format will be used throughout the book to define formal decision problems. The box contains the problem name, what we are given (the instance), and what is to be determined (the question). The decision question is always a yes-or-no question. (Technically, each decision problem corresponds to a formal language, consisting of the set of strings representing the instances for which the answer is "yes," under some encoding. See Appendix B for details.)

3.1   Zero-Player Games (Simulations)

We may think of deterministic, or zero-player, games as simulations: each move is determined from the preceding configuration. Examples that are often thought of as games include cellular automata, such as Conway's Game of Life [72]. The natural decision question for such games is whether a particular condition is ever satisfied in any configuration. For example, in the Game of Life, the question could be whether a particular cell is ever alive.

Constraint Logic.   The constraint-logic formalism does not restrict the set of moves available on a constraint graph to a unique next move from any given position. To consider a deterministic version, we must further constrain the legal moves. Rather than propose a rule that selects a unique next edge to reverse from each position, we apply determinism independently at each vertex, so that multiple edge reversals may occur on each deterministic "turn."

The basic idea is that each vertex should allow "signals" to "flow" through it if possible. So if both red edges reverse inward at an AND vertex, then on the next move the blue edge will reverse outward. For the bounded version, this idea is all we need. For the unbounded version, the rule is modified to allow inputs that cannot flow through to "bounce" back. (They cannot do so in the bounded version, because each edge can only reverse once.)

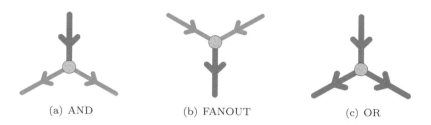

(a) AND   (b) FANOUT   (c) OR

Figure 3.1. Basis vertices for Bounded DCL.

3.1.1 Bounded Games

A bounded zero-player game is essentially a simulation that can only run for a linear time. It may seem a stretch to call such simulations "games," but they do fit naturally into the overall framework sketched in Figure I.1. Bounded Deterministic Constraint Logic is included here merely for completeness. Conceivably there could be some solitaire games where the player has no actual choice, and the game is of bounded length, but such games would not seem to be very interesting.

To formally define Bounded Deterministic Constraint Logic, we first define a constraint-graph successor operation. We define a vertex v as *firing* relative to an edge set R if its incident edges that are in R satisfy its minimum inflow, and $F(G, R)$ as the set of vertices in G that are firing relative to R. Then, if we begin with graph G_0 and edge set R_0,

$$R_{i+1} = \{e \mid e \text{ points to } v \text{ in } G_i, \ v \in F(G_i, R_0 \cup \ldots \cup R_i),$$
$$\text{and } e \notin R_0 \cup \ldots \cup R_i\},$$
$$G_{i+1} = G_i \text{ with edges in } R_{i+1} \text{ reversed.}$$

This process effectively propagates signals through a graph until they can no longer propagate.

> **BOUNDED DETERMINISTIC CONSTRAINT LOGIC (BOUNDED DCL)**
> **Instance:** Constraint graph G_0; edge set R_0; edge e in G_0.
> **Question:** Is there an i such that e is reversed in G_i?

Bounded DCL is P-complete. It remains P-complete when the graph G uses only the vertex types shown in Figure 3.1. However, unlike all the other constraint logics, Bounded DCL evidently does not remain P-complete when G is required to be planar. The proofs are given in Chapter 4.

3.1.2   Unbounded Games

Unbounded zero-player games are simulations that have no a priori bound on how long they may run. Cellular automata, such as Conway's Game of Life, are a good example.

Deterministic Constraint Logic (DCL) is the form of constraint logic that corresponds to these kinds of simulation. The definition is similar to the bounded case, removing the restriction that each edge may reverse at most once. However, this raises a problem: when an edge reverses into a vertex, the rule would have it reverse out again on the next step, as well as whatever other edges it enabled to reverse. This would lead to illegal configurations.

Therefore, we add the restriction that an edge that just reversed may not reverse again on the next step, unless on that step there are no other reversals away from the vertex to which the edge points. Again, we define a vertex v as firing relative to an edge set R if its incident edges that are in R satisfy its minimum inflow, and $F(G, R)$ as the set of vertices in G that are firing relative to R. Then, if we begin with graph G_0 and edge set R_0,

$$R_{i+1} = \{e \mid e \text{ points to } v \text{ in } G_i, \text{ and either } e \in R_i \text{ or } v \in F(G_i, R_i)$$
$$\text{but not both}\},$$
$$G_{i+1} = G_i \text{ with edges in } R_{i+1} \text{ reversed.}$$

The effect of this rule is that signals will flow through constraint graphs as desired, but when a signal reaches a vertex that it cannot "activate," or "flow through," it will instead "bounce." (See Figure C.1 in Appendix C for an example.)

This seems to be the most natural form of constraint logic that is unbounded and deterministic. It has the additional nice property that it is reversible. That is, if we start computing with G_{i-1} and R_i, instead of G_0 and R_0, we eventually get back to G_0.

> **DETERMINISTIC CONSTRAINT LOGIC (DCL)**
>
> **Instance:** Constraint graph G_0; edge set R_0; edge e in G_0.
>
> **Question:** Is there an i such that e is reversed in G_i?

DCL is PSPACE-complete. It remains PSPACE-complete when the graph G is required to be a planar graph that uses only the vertex types shown in Figure 3.2. The proofs are given in Chapter 4.

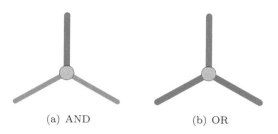

(a) AND   (b) OR

Figure 3.2. Basis vertices for DCL.

3.2   One-Player Games (Puzzles)

A one-player game is a puzzle: one player makes a series of moves, trying to accomplish some goal. For example, in a sliding-block puzzle, the goal could be to get a particular block to a particular location. We use the terms "puzzle" and "one-player game" interchangeably. For puzzles, the generic forced-win decision question—"does player X have a forced win?"— becomes "is this puzzle solvable?"

Constraint Logic. The one-player version of constraint logic is called Nondeterministic Constraint Logic (NCL). The rules are simply that on a turn the player may reverse any edge resulting in a legal configuration, and the decision question is whether a given edge may ever be reversed. For the bounded version, we allow each edge to reverse at most once.

Due to the simplicity of NCL, and the abundance of puzzles with reversible moves, it is often straightforward to find reductions showing various puzzles PSPACE-hard. This is the largest class of reductions presented in Part II.

3.2.1   Bounded Games

Bounded one-player games are puzzles in which there is a polynomial bound (typically linear) on the number of moves that can be made. Usually there is some resource that is used up. For example, in a game of Sudoku, the grid eventually fills up with numbers, and then either the puzzle is solved or it is not. In Peg Solitaire, each jump removes one peg, until eventually no more jumps can be made.

Bounded Nondeterministic Constraint Logic (Bounded NCL) abstracts the essence of a bounded puzzle. Bounded NCL is formally defined as follows:

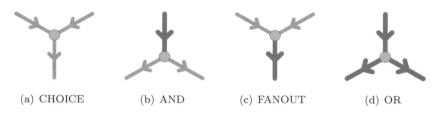

(a) CHOICE   (b) AND   (c) FANOUT   (d) OR

Figure 3.3. Basis vertices for Bounded NCL.

**BOUNDED NONDETERMINISTIC CONSTRAINT LOGIC
(BOUNDED NCL)**

Instance: Constraint graph G, edge e in G.

Question: Is there a sequence of moves on G that eventually
reverses e, such that each edge is reversed at most once?

Bounded NCL is NP-complete. It remains NP-complete when the graph
G is required to be a planar graph that uses only the vertex types shown
in Figure 3.3. It will also turn out to be useful to reduce from graphs that
have the property that only a single edge can initially reverse. Again, we
can assume this property in puzzle reductions; all these proofs are given in
Chapter 5.

3.2.2   Unbounded Games

Unbounded one-player games are puzzles in which there is no restriction on
the number of moves that can be made. Typically the moves are reversible.
For example, in a sliding-block puzzle, the pieces may be slid around in the
box indefinitely, and a block once slid can always be immediately slid back
to its previous position.

Nondeterministic Constraint Logic (NCL) is the form of constraint logic
that corresponds to this type of puzzle. It is formally defined as follows:

NONDETERMINISTIC CONSTRAINT LOGIC (NCL)

Instance: Constraint graph G, edge e in G.

Question: Is there a sequence of moves on G that eventually
reverses e?

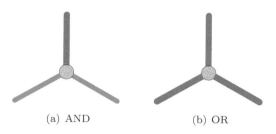

(a) AND   (b) OR

Figure 3.4. Basis vertices for NCL.

NCL is PSPACE-complete. It remains PSPACE-complete when the graph G is required to be a planar graph that uses only the vertex types shown in Figure 3.4 (basic AND and OR). The proofs are given in Chapter 5.

Protected-OR Graphs. We call an OR vertex *protected* if there are two of its edges that, due to global constraints, can never simultaneously be directed inward. Intuitively, graphs with only protected ORs are easier to reduce to another problem domain, since the corresponding OR gadgets need not function correctly in all the cases that a true OR must. We will show that we can assume only protected ORs in puzzle reductions.

Configuration-to-Configuration Problem. Occasionally it will be desirable to reduce NCL to a problem in which the goal is to achieve some particular total state, rather than an isolated partial state. For example, in many sliding-block puzzles the goal is to move a particular piece to a particular place, but in others it is to reach some given complete configuration. (One version of such a problem is the Warehouseman's Problem (Section 9.4)).

For attacking such problems, we show that an additional variant of NCL is hard: the goal is to achieve a given total graph state, rather than an individual edge reversal.

3.3 Two-Player Games

With two-player games, we are finally in territory familiar to classical game theory and combinatorial game theory. Two-player, perfect-information games are also the richest source of existing hardness results for games. In a two-player game, players alternate making moves, each trying to achieve some particular objective. The standard decision question is "does player X have a forced win from this position?"

Constraint Logic. The two-player version of constraint logic, Two-Player Constraint Logic (2CL), is defined as might be expected. To create different moves for the two players, Black and White, we label each constraint graph edge as either Black or White. (This is independent of the red/blue coloration, which is simply a shorthand for edge weight.) Black (White) is allowed to reverse only Black (White) edges on his move. Each player has a target edge that he is trying to reverse.

3.3.1 Bounded Games

Bounded two-player games are games in which there is a polynomial bound (typically linear) on the number of moves that can be made. As with bounded puzzles, usually there is some resource that is used up. In Hex, for example, each move fills a space on the board, and when all the spaces are full, the game must be over. Similarly, in Amazons, on each move an amazon must remove one of the spaces from the board. In Konane, each move removes at least one stone. There are many other examples of bounded two-player games. When the resource is exhausted, the game cannot continue.

Bounded Two-Player Constraint Logic is the natural form of constraint logic that corresponds to this type of game. It is formally defined as follows:[1]

> **BOUNDED TWO-PLAYER CONSTRAINT LOGIC (BOUNDED 2CL)**
>
> **Instance:** Constraint graph G, partition of the edges of G into sets B and W, and edges $e_B \in B$, $e_W \in W$.
>
> **Question:** Does White have a forced win in the following game? Players White and Black alternately make moves on G. White (Black) may only reverse edges in W (B). Each edge may be reversed at most once. White (Black) wins (and the game ends) if he ever reverses e_W (e_B).

One remark about this definition is in order. In the field of combinatorial game theory, it is normal to define the loser as the first player unable to move. Games are thus about maximizing one's number of available moves. This definition would work perfectly well for 2CL, rather than using target edges to determine the winner; the hardness reduction would not be substantially altered, and the definition would seem to be a bit more concise.

---

[1] We can assume without loss of generality that the game ends with no winner if a player is unable to move.

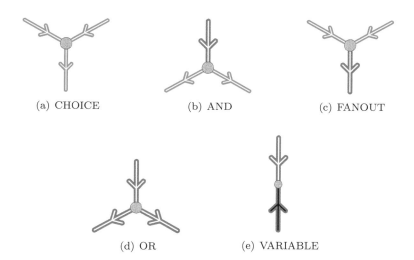

(a) CHOICE   (b) AND   (c) FANOUT

(d) OR   (e) VARIABLE

Figure 3.5. Basis vertices for Bounded 2CL.

However, the definition above is more consistent with the other varieties of constraint logic. Always, the goal is to reverse a given edge.

Bounded 2CL is PSPACE-complete. It remains PSPACE-complete when the graph G is required to be a planar graph that uses only the vertex types shown in Figure 3.5. The proofs are given in Chapter 6.

3.3.2 Unbounded Games

Unbounded two-player games are games in which there is no restriction on the number of moves that can be made. Typically (but not always) the moves are reversible. Examples include the classic games Chess, Checkers, and Go.

Two-Player Constraint Logic (2CL) is the form of constraint logic that corresponds to this type of game. It is formally defined as follows:

TWO-PLAYER CONSTRAINT LOGIC (2CL)

Instance: Constraint graph G, partition of the edges of G into sets B and W, and edges $e_B \in B$, $e_W \in W$.

Question: Does White have a forced win in the following game? Players White and Black alternately make moves on G. White (Black) may only reverse edges in W (B). White (Black) wins if he ever reverses e_W (e_B).

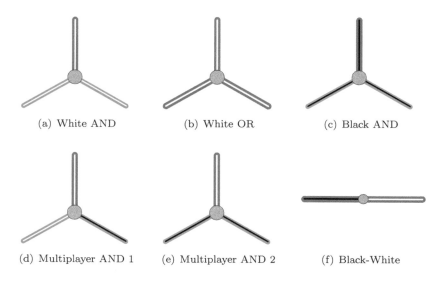

(a) White AND   (b) White OR   (c) Black AND

(d) Multiplayer AND 1   (e) Multiplayer AND 2   (f) Black-White

Figure 3.6. Basis vertices for 2CL.

The 2CL problem is EXPTIME-complete. It remains EXPTIME-complete when the graph G is required to be a planar graph that uses only the vertex types shown in Figure 3.6. (Note that color symmetries in a target problem could mean that the Black AND gadget would be a color-swapped version of the White one, so there could be only five gadgets to build for a reduction.) The proofs are given in Chapter 6.

3.4 Team Games

It turns out that adding players beyond two to a game does not increase the complexity of the standard decision question, "does player X have a forced win?" We might as well assume that all the other players team up to beat X, in which case we effectively have a two-player game again. If we generalize the notion of the decision question somewhat, we do obtain new kinds of games. In a *team game*, there are still two "sides," but each side can have multiple players, and the decision question is whether *team* X has a forced win. A team wins if any of its players wins.

Team games with perfect information are still just two-player games in disguise, however, because again all the players on a team can cooperate and play as if they were a single player. However, when there is hidden information, then team games turn out to be different from two-player games. (We could think of a team in this case as a player with a peculiar kind of mental limitation—on alternate turns he forgets some aspects of his situation, and remembers others.) Therefore, we will only consider team games of imperfect information, and we will sometimes simply refer to them simply as "team games."

Constraint Logic. The natural team, private-information version of constraint logic assigns to each player a set of edges he can control, and a set of edges whose orientation he can see. As always, each player has a target edge he must reverse to win. To enable a simpler reduction to the unbounded form of team constraint logic, we allow each player to reverse up to some given constant k edges on his turn, rather than just one, and leave the case of $k = 1$ as an open problem.

3.4.1 Bounded Games

Bounded team games of imperfect information include card games such as Bridge. Here we can consider one hand to be a game, with the goal being either to make the bid, or, if on defense, to set the other team. The hand is of a bounded size, so the game must end in a bounded number of moves. Focusing on a given hand also removes the random element from the game, making it potentially suitable for study within the present framework.

We define the corresponding type of constraint logic, Bounded Team Private Constraint Logic (Bounded TPCL), as follows:

BOUNDED TEAM PRIVATE CONSTRAINT LOGIC (BOUNDED TPCL)

Instance: Constraint graph G; integer N; for $i \in \{1 \dots N\}$: sets $E_i \subset V_i \subset G$, edges $e_i \in E_i$; partition of $\{1 \dots N\}$ into nonempty sets W and B.

Question: Does White have a forced win in the following game? Players $1 \dots N$ take turns in that order. Player i only sees the orientation of the edges in V_i, and moves by reversing an edge in E_i that has not previously reversed; a move must be known legal based on V_i. White (Black) wins if Player $i \in W$ (B) ever reverses edge e_i.

Bounded TPCL is NEXPTIME-complete. It remains NEXPTIME-complete when the graph G is required to be planar and use only AND and OR vertices. (Note that in this case there are many different subtypes of AND and OR vertices.) The proof is given in Chapter 7.

3.4.2   Unbounded Games

In general, team games of private information are undecidable. This result
was claimed by Peterson and Reif in 1979 [133]. However, there are a few
problems with the proof, which we address in Section 7.2. Strangely, the
result also seems to be not very well known. Part of the problem may be
that the authors seem to consider the result of secondary importance to
the other results in [133]. From our perspective, however, the fact that
there are undecidable space-bounded games is fundamental to the viewpoint that games are an interesting model of computation. It both shows
that games are as powerful as general Turing machines, and highlights the
essential difference from the Turing-machine foundation of theoretical computer science, namely that *a game computation is a manipulation of finite
resources*. Thus, this seems to be a result of some significance.

It might seem that the concept of an unbounded-length team game of
private information is getting rather far from the intuitive notion of game.
However, individually each of these attributes is common in games. There
is at least one actual game that fits this category, called Rengo Kriegspiel.
This is a team, blindfold version of Go. (See Appendix A for details.)
One of the authors has personally played this game on a few occasions,
and it is intriguing to think that it's possible he has played the hardest
game in the world, which cannot even in principle be played perfectly by
any algorithm.

Team Private Constraint Logic is defined in the box that follows. Note
the addition of the parameter k relative to the bounded case. This is,
admittedly, an extra generalization to make a reduction easier; nonetheless,
it is a reasonable generalization, and all other constraint-logic games in this
book are naturally restricted versions of this game.

TEAM PRIVATE CONSTRAINT LOGIC (TPCL)

Instance: Constraint graph G; integer N; for $i \in \{1 \dots N\}$:
sets $E_i \subset V_i \subset G$, edges $e_i \in E_i$; partition of $1 \dots N$ into
nonempty sets W and B; integer k.

Question: Does White have a forced win in the following game?
Players $1 \dots N$ take turns in that order. Player i only sees
the orientation of the edges in V_i, and moves by reversing
up to k edges in E_i; a move must be known legal based on
V_i. White (Black) wins if Player $i \in W$ (B) ever reverses
edge e_i.

TPCL is undecidable. It remains undecidable when the graph G is
required to be planar and use only AND and OR vertices. (Note that in this

case there are many different subtypes of AND and OR vertices.) The proof
is given in Chapter 7.

Zero-Player Games (Simulations)

In this chapter we present the definitions and complexity proofs for Deterministic Constraint Logic, both the bounded and unbounded varieties. We also briefly discuss some potential real-world applications of Deterministic Constraint Logic for building reversible computers.

Complexity Background.   We may think of deterministic, or zero-player, games as simulations: each move is determined from the preceding configuration. Examples that are often thought of as games include cellular automata, such as Conway's Game of Life [72]. The natural decision question for such games is whether a particular condition is ever satisfied in any configuration. For example, in the Game of Life, the question could be whether a particular cell is ever alive.

More generally, the class of zero-player games corresponds naturally to ordinary computers, or deterministic space-bounded Turing machines—the kinds of computation tools we have available in the real world, at least until quantum computers develop further.

PSPACE is the class of problems that can be solved by a deterministic space-bounded Turing machine. Unsurprisingly, Life is PSPACE-complete. Actually, Life has been shown to be "computation universal" [8,141,170] on an infinite grid; that is, it can simulate an arbitrary Turing machine. That means that there are decision questions about a Life game (e.g., "will this cell ever be born") that are undecidable. On a finite grid, the corresponding property is PSPACE-completeness. (This result is not mentioned explicitly in the cited works, but it does follow directly, at least from [141].)

For bounded games, the corresponding model of computation is the deterministic time-bounded Turing machine, and the appropriate complexity class is P.

Constraint Logic. The constraint-logic formalism does not restrict the set of moves available on a constraint graph to a unique next move from any given position. To consider a deterministic version, we must further constrain the legal moves. Rather than propose a rule that selects a unique next edge to reverse from each position, we apply determinism independently at each vertex, so that multiple edge reversals may occur on each deterministic "turn."

The basic idea is that each vertex should allow "signals" to "flow" through it if possible. So if both red edges reverse inward at an AND vertex, then on the next move the blue edge will reverse outward. For the bounded version, this idea is all we need. For the unbounded version, the rule is modified to allow inputs that cannot flow through to "bounce" back. (They cannot do so in the bounded version, because each edge can only reverse once.)

4.1   Bounded Games

A bounded zero-player game is essentially a simulation that can only run for a linear time. It may seem a stretch to call such simulations "games," but they do fit naturally into the overall framework sketched in Figure I.1. Bounded Deterministic Constraint Logic is included here merely for completeness. Conceivably there could be some solitaire games where the player has no actual choice, and the game is of bounded length, but such games would not seem to be very interesting.

Deciding such games can clearly be done in polynomial time—just run the simulation and check the result. In general, such games can also be P-hard.

To formally define Bounded Deterministic Constraint Logic, we first define a constraint-graph successor operation. We define a vertex v as *firing* relative to an edge set R if its incident edges that are in R satisfy its minimum inflow, and $F(G, R)$ as the set of vertices in G that are firing relative to R. Then, if we begin with graph G_0 and edge set R_0,

$$R_{i+1} = \{e \mid e \text{ points to } v \text{ in } G_i,\ v \in F(G_i, R_0 \cup \ldots \cup R_i),$$
$$\text{and } e \notin R_0 \cup \ldots \cup R_i\},$$
$$G_{i+1} = G_i \text{ with edges in } R_{i+1} \text{ reversed.}$$

This process effectively propagates signals through a graph until they can no longer propagate.

BOUNDED DETERMINISTIC CONSTRAINT LOGIC (BOUNDED DCL)
Instance: Constraint graph G_0; edge set R_0; edge e in G_0.
Question: Is there an i such that e is reversed in G_i?

4.1.1   P-completeness

This style of computation is captured by the notion of *Boolean circuits*, and more specifically, *monotone Boolean circuits*. A monotone Boolean circuit is a directed acyclic graph where the nodes are *gates* (*AND* or *OR*) or *inputs*, and the connections and edge orientations are as expected for the gate types. The input nodes are either *true* or *false*. The gates are allowed to have multiple outputs (= outward directed edges); that is, there is fanout. One gate is the *output gate*. Each gate computes the appropriate Boolean function of its input. The *value* of the circuit, for a given assignment of Boolean input values, is the value computed by the output gate. An ordinary (non-monotone) Boolean circuit is also allowed NOT gates; these turn out not to add any computational power.

Essentially, then, a monotone Boolean circuit is just a representation of a monotone Boolean formula, that potentially allows some space savings by reusing subexpressions via fanout. The problem of determining the value of a monotone Boolean circuit, called Monotone Circuit Value, is P-complete [77]. We show that Bounded DCL is P-complete by a reduction from Monotone Circuit Value.

Theorem 4.1. *Bounded DCL is P-complete.*

Proof: Given a Boolean circuit C, we construct a corresponding Bounded DCL problem, such that the edge e in the DCL problem is reversed just when the circuit value is true. This process is straightforward: for every gate in C we create a corresponding vertex, either an AND or an OR. When a gate has more than one output, we use AND vertices in the FANOUT configuration. The difference here between AND and FANOUT is merely in the initial edge orientation. For inputs, we use free edges: true inputs are directed into their gates and are included in R_0; false inputs are directed away from their gates.

Then, the Bounded DCL dynamics exactly mirror the operation of the Boolean circuit, and e will eventually reverse if and only if the circuit value is true. This shows that Bounded DCL is P-hard. Clearly it is also in P: we may compute G_{i+1} from G_i in linear time (keeping track of which

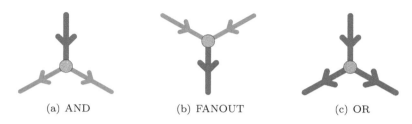

(a) AND   (b) FANOUT   (c) OR

Figure 4.1. Basis vertices for Bounded DCL.

edges have already reversed), and after a linear time no more edges can
ever reverse.   □

4.1.2 Restricted Problem

For all of the other kinds of constraint-logic games, it turns out that restricting the constraint graphs to planar configurations does not change
their computational power. However, planar Bounded DCL seems to be
weaker than unrestricted Bounded DCL. The reason is that, while Monotone Circuit Value is P-complete, the planar monotone circuit value problem has been shown to lie in NC^3 [174], and it is believed that $NC^3 \subsetneq P$.
Planarity is a useful property to have in constraint logic, because it greatly
simplifies reductions to other games. (In this case, however, there are no
obvious games one would be interested in showing P-complete anyway, or
if there are they have escaped our notice.)

However, we can strengthen Theorem 4.1 to apply to graphs using a
restricted set of vertices. Then, for reductions from Bounded DCL to other
games, it would be sufficient to build this restricted set of gadgets.

Theorem 4.2. *Bounded DCL is P-complete, even for graphs using only the
vertex types shown in Figure 4.1.*

Proof: The reduction from Monotone Circuit Value uses only the vertices
shown in Figure 4.1, plus red-blue vertices (to join OR outputs to AND
inputs, or FANOUT outputs to OR inputs), and free edges for the inputs.
We use the conversion techniques in Section 2.3 to convert these to AND
and OR vertices.[1]   □

---

[1] Some of the required edge orientations do not correspond to the initial orientations
of the basis vertices in Figure 4.1. They do, however, correspond to legal states of
the basis vertices; we assume that some of the edges have "already reversed," and are
included in R_0. Reductions from Bounded DCL are unaffected by this subtlety.

4.2   Unbounded Games

Unbounded zero-player games are simulations that have no a priori bound on how long they may run. Cellular automata, such as Conway's Game of Life, are a good example.

Such games can always be decided in polynomial space—PSPACE—by simply running the simulation long enough so that its state must begin to loop. Since there are only finitely many states, and each step is deterministic, this must happen eventually. The time required is exponential in the space used to represent a configuration, but the space required to store a counter is only polynomial.

Deterministic Constraint Logic (DCL) is the form of constraint logic that corresponds to these kinds of simulation. The definition is similar to the bounded case, removing the restriction that each edge may reverse at most once. However, this raises a problem: when an edge reverses into a vertex, the rule would have it reverse out again on the next step, as well as whatever other edges it enabled to reverse. This would lead to illegal configurations.

Therefore, we add the restriction that an edge that just reversed may not reverse again on the next step, unless on that step there are no other reversals away from the vertex to which that edge points. Again, we define a vertex v as firing relative to an edge set R if its incident edges that are in R satisfy its minimum inflow, and $F(G, R)$ as the set of vertices in G that are firing relative to R. Then, if we begin with graph G_0 and edge set R_0,

$$R_{i+1} = \{e \mid e \text{ points to } v \text{ in } G_i, \text{ and either } e \in R_i \text{ or } v \in F(G_i, R_i)$$
$$\text{but not both}\},$$
$$G_{i+1} = G_i \text{ with edges in } R_{i+1} \text{ reversed.}$$

The effect of this rule is that signals will flow through constraint graphs as desired, but when a signal reaches a vertex that it cannot "activate," or "flow through," it will instead "bounce." (See Figure C.1 in Appendix C for an example.) In DCL figures, edges that have just reversed are highlighted; they are a relevant part of the state.

This seems to be the most natural form of constraint logic that is unbounded and deterministic. It has the additional nice property that it is reversible. That is, if we start computing with G_{i-1} and R_i, instead of G_0 and R_0, we eventually get back to G_0.

DETERMINISTIC CONSTRAINT LOGIC (DCL)

Instance: Constraint graph G_0; edge set R_0; edge e in G_0.
Question: Is there an i such that e is reversed in G_i?

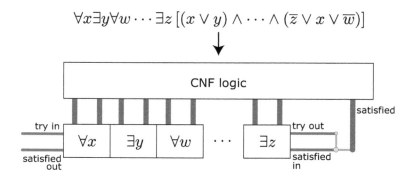

Figure 4.2. Schematic of reduction from QBF.

4.2.1   PSPACE-completeness

We show that DCL is PSPACE-complete via a reduction from Quantified Boolean Formulas (QBF; see Section B.5.3). Given an instance of QBF (a quantified Boolean formula F in CNF), we construct a corresponding constraint graph G such that iteration of the above deterministic rule will eventually reverse a target edge e if and only if F is true. The reduction is shown schematically in Figure 4.2.

This reduction is not intended as a suggestion for how to actually perform computations with such circuits, were they to be physically realized. This is because that reduction entails an exponential slowdown from the computation performed on the corresponding space-bounded Turing machine. Instead, we later show how a practical reversible computer could be built using either dual-rail logic or Fredkin gates made with DCL components. Those constructions require the addition of a few (nonreversible, entropy-generating) elements that do not strictly fit within the DCL model, however, so they are not sufficient for showing PSPACE-completeness.

The QBF reduction is rather elaborate, and many details are relegated to Appendix C. While this chapter is logically the first to deal with specific forms of constraint logic, the corresponding reduction in Chapter 5 is similar but more straightforward, and we suggest that the reader skip this reduction until after reading that one.

Reduction.   One way to determine the truth of a quantified Boolean formula is as follows: Consider the initial quantifier in the formula. Assign its variable first to false and then to true, and for each assignment, recursively ask whether the remaining formula is true under that assignment. For an existential quantifier, return true if either assignment succeeds; for a universal quantifier, return true only if both assignments succeed. For the base case, all variables are assigned, and we only need to test whether the CNF formula is true under the current assignment.

The constructed constraint graph G operates in a similar fashion. Initially, the **try in** edge reverses into the left quantifier gadget, activating it. When a quantifier gadget is activated, it tries both possible truth values for the corresponding variable. For each, it sends the appropriate truth value into the CNF logic circuitry. The CNF circuitry then sends a signal back to the quantifier gadget, having stored the value. The quantifier then activates the next quantifier's **try in** edge.

Eventually the last quantifier will set an assignment. Then, if the formula is satisfied by this total assignment, the last quantifier's **satisfied in** edge will activate. When a quantifier receives a **satisfied in** signal, if it is an existential quantifier, then it simply passes it back to the previous quantifier: the assignment has succeeded. For a universal quantifier, when the first assignment succeeds, an internal memory bit is set. When the second assignment succeeds, if the memory bit is set, then the quantifier activates the previous quantifier's **satisfied in** input.

The leftmost **satisfied out** edge will eventually reverse if and only if the formula is true.

Timing. Since multiple signals can be bouncing around simultaneously in a DCL graph, and signals must arrive "in phase" to activate an AND vertex, timing issues are critical when designing DCL gadgets. To simplify analysis, all gadget inputs and outputs will be assumed to occur at times 0 mod 4. By this we mean that after an input has arrived, the first propagating edge reversal inside the gadget will be at time 1 mod 4, and the last edge reversal inside the gadget before the signal propagates out will be at time 0 mod 4.

Timing issues cause us to use blue-blue and red-red vertices in our reduction; these vertex types are not needed for other kinds of constraint logic. Essentially, these vertices are simply wires allowing a signal to flow from one edge to the other; we need these delay wires to create the correct signal phases. (Later we show that we can translate the constructions into ANDs and ORs.)

Quantifier Gadgets. The existential- and universal-quantifier gadgets are shown in Figure 4.3. Their properties will be stated here; the simplest proof that they operate as described is simply a series of snapshots of states, each of which clearly follows from the previous by application of the deterministic rule. These snapshots are given in Appendix C. (All circuits were designed and tested with a DCL circuit simulator.) Note that edges that might appear to be extraneous serve to synchronize output phases, as described above. Each quantifier's **try out** edge is connected via a blue-blue vertex to the next quantifier's **try in edge** (except for the last quantifier,

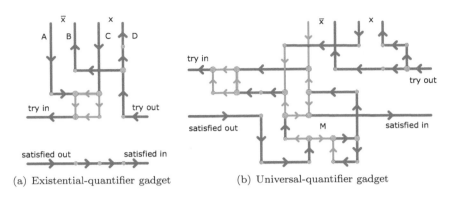

(a) Existential-quantifier gadget   (b) Universal-quantifier gadget

Figure 4.3. DCL quantifier gadgets.

described later); similarly for **satisfied in** and **satisfied out**. The x and $\overline{x}$ edges likewise connect to the CNF logic circuitry, described later.

An existential quantifier assigns its variable to be first false, and then true. If either assignment succeeds, the quantifier activates its **satisfied out** edge. The simplest switching circuit that performs this task also assigns its variable to false again one more time, but this does not alter the result.

The gadget is activated when its **try in** edge reverses inward (at some time 1 mod 4). The internal red edges cause the signal to propagate to edge **A** three steps later. The signal then proceeds into CNF logic circuitry, described later, and returns via edge **B**. It then propagates to **try out** three steps later. Now, it is possible that **satisfied in** will later reverse inward; if so, **satisfied out** then reverses three steps later. Then, later, **satisfied out** may reverse inward, and **satisfied in** will then reverse outward three steps later. Here the sense of input and output is being reversed—the performed operation is being "undone."

Regardless of whether **satisfied in** was activated, later **try out** will reverse back inward, continuing to unwind the computation. Then **B** will reverse again three steps later, and eventually **A** will reverse inward. Then, the switching circuit composed of the red edges will send the signal to **C** three steps later, effectively setting the variable to be true. Then, the same sequence as before happens, except that edges **C** and **D** are used instead of **A** and **B**. Finally, the switching circuit tries **A** and **B** again, and then at last **try in** is directed back outward. The switching operates based on stored internal state in the red edges; see Appendix C for details.

The universal-quantifier gadget is similar, but more complicated. It uses the same switching circuit to assign its variable first to false and then to true, and then to false once more. However, if the false assignment succeeds—that is, if **satisfied in** is directed inward when the variable is set to false—then instead of propagating the signal back to the previous quantifier, the success is remembered by reversing some internal edges so that edge M is set to reversing every step. Then, if **satisfied in** is later directed in while M is in this state, and the variable is set to true, these conditions cause **satisfied out** to direct outward. Finally, in this case setting the variable to false again is useful; this causes M and the other internal edges to be restored to their original state, erasing the memory of the success when setting the variable false. Then, again, **try in** is directed back outward; the gadget has cleaned up and deactivated, waiting for the next assignment of the leftward variables.

CNF Logic. We already have AND and OR vertices, so it might seem that we could simply feed the variable outputs from the quantifiers into a network of these that corresponds to the Boolean formula, and its output would activate only when the formula was satisfied by that assignment. However, the variable signals would all have to arrive simultaneously for that approach to work. Furthermore, ANDs that had only one active input would bounce that signal back, potentially confusing the timing in the gadget that sent the signal.

Rather than try to solve all such timing issues globally, we follow a strategy of keeping a single path of activation, that threads its way through the entire graph. Each gadget need merely follow the phase timing constraints described above.

We build abstract logic gates, AND′, OR′, and FANOUT′, that operate differently from their single-vertex counterparts. These gates receive inputs, remember them, and acknowledge receipt by sending a return signal. If appropriate, they also send an output signal, in a similar paired fashion, prior to sending the return signal. Later, the input signals are turned off in the reverse order that they were turned on, by sending a signal back along the original return signal line; the gadgets then complete the deactivation by sending the signal back along the original input activation line.

This description will be made clearer by seeing some examples. As with the quantifier gadgets, correct operation of the CNF gadgets is demonstrated in the snapshots in Appendix C. These gadgets are connected together to correspond to the structure of the CNF formula: variable outputs feed OR′ inputs; OR′ outputs feed other OR′ inputs or AND′ inputs; AND′ outputs feed other AND′ inputs, except for the final formula output, which is combined with the final quantifier output as described later.

The AND′ gadget is shown in Figure 4.4(a). We assume that, if both inputs will arrive and the gate will activate, **input 1** will arrive first; later we justify this assumption. Suppose a signal arrives at input 1, on edge A. Then, as in the universal-quantifier gadget, edge M will be set bouncing to remember that this input has been asserted. If input 2 later activates, along edge C, then the same switching circuit as used in the quantifiers will send a signal so that it will arrive in phase with M, and activate the output on edge E. Later, when acknowledgment of this signal is received on edge F, the return signal is propagated back to the second input via D.

Suppose, instead, that a signal arrives at C when one has not first arrived at A. By assumption, the variable feeding into A is then false, so the AND′ should not activate. In this case the internal switch gadget sends a signal toward E, but because M is not in the right state it bounces back, and is then switched to the exit at D. Thus, the gate has acknowledged the second input, without activating the output. The reverse process occurs when D is redirected in.

The OR′ gadget, shown in Figure 4.4(b), is significantly more complicated. This is because the gate must robustly handle the case where one input arrives and activates the output, and later the other input arrives. The gate needs to know that it has already been activated, and simply reply to the second activation gracefully. (Note the highlighted edges in Figure 4.4(b); these are edges that have just reversed, and thus would be in the input edge set R_0.)

If an input arrives on edge A, the left switch gadget directs it up to the central OR vertex O, and then on to the output edge E. When the return signal arrives via F, the upper switch gadget S tries first one side and then the other. The edge left bouncing at M is in phase with this signal, which then propagates to B. The corresponding edge N is not bouncing, so when the signal from S arrives there it bounces back. Switch S has extra edges relative to the other switches; these create the proper signal phases. The entire process reverses when the signal is returned through B, first turning off the output, then returning via A. Since the gate is symmetric, a single input arriving at C first also behaves just as described, sending its return signal along D.

Suppose an input arrives at C after one has arrived at A and left at B, that is, when the gate is "on." Then when the signal reaches the OR vertex O it will propagate on toward M, and not toward E (because the path to E is already directed outward). But M will be directed away at this point, and the signal will bounce back, finally exiting at D with the right phase. Again, when the signal returns via D the reverse process sends it back to C. All of these sequences are shown explicitly in Appendix C.

The FANOUT′ gadget, shown in Figure 4.5(a), is straightforward. An input arriving on edge A is sent to outputs 1 and 2 in turn; then the return signal is sent back on edge B. The reverse sequence inactivates the gadget.

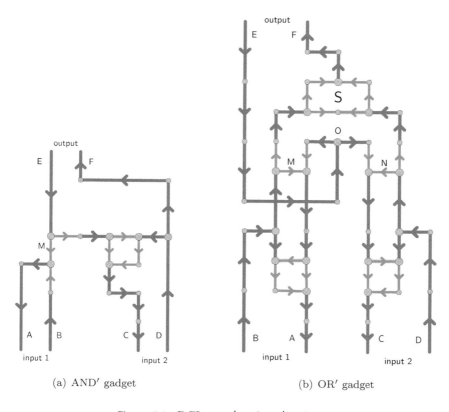

(a) AND′ gadget

(b) OR′ gadget

Figure 4.4.   DCL AND′ and OR′ gadgets.

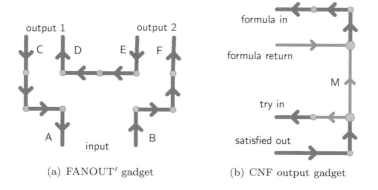

(a) FANOUT′ gadget

(b) CNF output gadget

Figure 4.5.   Additional CNF gadgets.

Remaining Connections. To start the computation, we attach a free edge terminator, as shown in Figure 2.5(a), to the leftmost quantifier's **try in** edge, and set that edge to reversing from G_0 to G_1. Similarly, we attach another free edge terminator to the leftmost quantifier's **satisfied out** edge; this is the target edge e that will reverse just when the QBF formula is true.

Finally, we must have a way to connect the rightmost quantifier and CNF outputs together, and feed the result into the rightmost quantifier's **satisfied in** input. This is done with the graph shown in Figure 4.5(b). The rightmost quantifier's **try out** edge connects to the **try in** edge here, and its **satisfied in** edge connects to the **satisfied out** edge here. The output of the CNF formula's final AND′ gadget connects to **formula in**, and its return output edge connects to **formula return**.

If the formula is ever satisfied by the currently activated variable assignment from all the quantifiers, then a signal will arrive at **formula in** and exit at **formula return**, leaving edge M bouncing every step. Then, when the last quantifier activates its **try out**, the signal arriving at **try in** will be in phase with M, propagating the signal on to the last quantifier's **satisfied in** input, where it will be processed as described above. If the formula is not satisfied, M will still be pointing up, and the **try in** signal will just bounce back into the last quantifier.

AND′ Ordering. As mentioned above, we assume that **input 1** of an AND′ will always activate, if at all, before **input 2**. However, in a general quantified CNF formula, it is not the case that the clauses need be satisfied in any predetermined order, if the variables are assigned in the order quantified. To solve this problem, we modify the circuit described above as follows. We protect **input 2** of every AND′ a_1 in the original circuit with an additional AND′ a_2, so that the original **input 2** signal now connects to a_2's **input 1**, and a_2's **output** connects to a_1's **input 2**.

Then, the rightmost quantifier's **try out** edge, instead of connecting directly to the merge gadget shown in Figure 4.5(b), first threads through every newly introduced AND′ **input 2** pathway, and then from there connects to the merge gadget. (We make sure that the introduced pathways are the right lengths to satisfy the timing rule.) Thus, as the variable values are set, when an AND′ would have its second input arrive before the first in the original circuit, in the modified circuit the second input's activation is deferred until its first input has had a chance to arrive. This way, we can ensure that the inputs arrive in the right order, by threading the path from the final **try out** appropriately.

Theorem 4.3. *DCL is PSPACE-complete.*

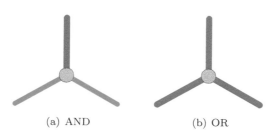

(a) AND   (b) OR

Figure 4.6. Basis vertices for DCL.

Proof: Given a quantified Boolean formula F, we construct DCL graph G_0 and edge set R_0 as described above. The individual gadgets' correct operation is explicitly shown in Appendix C. The leftmost quantifier's satisfied out edge will eventually reverse if and only if F is true. This shows that DCL is PSPACE-hard.

DCL is also clearly in PSPACE: a simple, constant-space deterministic algorithm executes each deterministic step of the graph, and detects when the target edge reverses.   □

4.2.2   Restricted Problem

For reductions from DCL to other simulations, we need to strengthen Theorem 4.3 to apply to planar graphs that use a restricted set of vertex types. (We do not present any DCL reductions in this book; however, an application of DCL to evolutionary graph theory is currently in preparation.)

Theorem 4.4. *DCL is PSPACE-complete, even for planar graphs using only the vertex types shown in Figure 4.6.*

Proof: First, we convert the constructed graph G in the above reduction to an equivalent graph G' that uses only ANDs and ORs, as follows. Replace every edge with a sequence of four new edges: each blue edge is replaced by a chain of four blue edges; each red edge is replaced with a chain of four edges colored red, blue, blue, and red. However, wherever there is a red-red vertex in G, use a blue edge at those endpoints of the new chains, rather than red. That is, red-red becomes a chain red, blue, blue, blue, blue, blue, blue, red. The new graph has blue-blue and red-blue vertices, but no red-red vertices. For the blue-blue vertices, use the technique of Section 2.3: attach a constrained loose blue edge. This edge will permanently point away from the vertex, and thus the behavior will be identical to that at an actual blue-blue vertex. For the red-blue vertices, again use

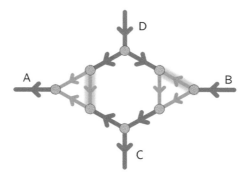

Figure 4.7. DCL crossover gadget.

the subgraph given in Section 2.3. Here, we must be careful with timing. The two red edges that provide the extra inputs to the red-blue vertices (see Figure 2.4) will "bounce" each turn, as long as the blue edge is directed inward. However, we can easily arrange for the phase of the bounce to be such that whenever a signal arrives on the incoming red edge, the extra edge will also point into the vertex. This is because such reversals must always occur on a global odd time step.

Next we must make the graph planar. The gadget shown in Figure 4.7 (which is equivalent to the normal "half-crossover" gadget (Figure 2.2(b)), with appropriate initial orientations and phases) shows how to cross signals in DCL. (The gadget must be padded with extra edges to satisfy the timing constraints.) As shown in Appendix C, a signal arriving at edge **A** will exit via **B**, and likewise for **C** and **D**. The sequence **A** to **B**, **C** to **D** also works, as do reverses of all these sequences. However, one sequence that does not work as desired is the following: **C** to **D**, **A** to **B**. After a **C** to **D** traversal, a signal arriving at **A** exits at **C** instead of **B**.

This limitation will not matter for our application, however; all crossings in the above construction are of the form that one edge will always be activated first, and the second, if activated, will deactivate before the first. This may be verified for the crossings within the gadgets by examining the activation sequences in Appendix C. For the other crossings, within the CNF logic, the global pattern of activation ensures that a pathway is never deactivated until any pathways it has activated are first deactivated.

To see that a particular crossover input always arrives first when two eventually arrive within the CNF logic, note that there are two types of such crossings: variable outputs crossing on their way to OR' inputs, and extra crossings created by the AND'-ordering pathway discussed above. (The OR' outputs do not need to cross on their way to the AND' inputs, because in

CNF this part of the network is just a tree.) In the first case, whenever both crossover inputs are active, the one from the earlier variable in the quantification sequence clearly must have arrived first. In the second case, the crossover input from the AND′-ordering pathway must always arrive after the input from the path it is crossing.

Everywhere edges cross in the original construction, we replace that crossing pair by a crossover gadget, suitably padded with extra edges to satisfy the timing requirements. We can easily ensure that it takes time 1 mod 4 to traverse a crossover in either direction; this guarantees that the gadget timing will be insensitive to the replacement.   □

4.2.3   Efficient Reversible Computation

Ordinary computers are not reversible. As a result, the information losses that are constantly occurring in an ordinary computer result in an increase in entropy, manifested as heat. This is known as *Landauer's Principle*: every bit of lost information results in a dissipation of $kT \ln 2$ joules of energy [112] (where k is Boltzmann's constant). Thus, as computers perform more operations per second, they require more power, and dissipate more heat. This seems obvious. However, it is theoretically possible to build a reversible computer, in which all of the atomic steps, except for recording desired output, are reversible [6], and dissipate arbitrarily little heat. Reversible computing is an active area of research (see, e.g., [61]), with many engineering challenges, and potential for dramatic increases in effective computing power over conventional computers.

Deterministic Constraint Logic represents a new style of reversible computation, which could potentially have practical application. It could be that the DCL deterministic edge-reversal rule is possible to implement effectively on a microscopic, perhaps molecular, level; certainly, the basic mechanism of switching edges between two states is suggestive of a simple physical system with two energy states.

As mentioned earlier, the construction showing DCL PSPACE-complete is not useful from a practical standpoint, because it involves an exponential slowdown in the reduction from Turing machine to QBF formula to DCL process. However, it is possible to build conventional kinds of reversible computing elements from DCL components. Figure 4.8(a) shows a Fredkin gate [63]. This gate has the property that a signal arriving on edge A, B, or C will be propagated to D, E, or F, respectively, but A and B in combination will activate D and F, and A and C in combination will activate D and E. Effectively, it is a "controlled switch" gate. It is also possible to directly implement a reversible dual-rail logic, as shown in Figure 4.8(b). Real computers built from such DCL gadgets would still require some localized nonreversible components, as would any usable reversible computer, which

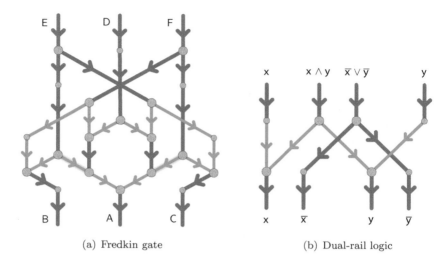

(a) Fredkin gate   (b) Dual-rail logic

Figure 4.8. DCL reversible computation gadgets.

is why these gadgets were not the basis for the PSPACE-completeness
proof.

One-Player Games (Puzzles)

In this chapter we present the definitions and complexity proofs for Nondeterministic Constraint Logic, both the bounded and unbounded varieties.

Complexity Background. A one-player game is a puzzle: one player makes a series of moves, trying to accomplish some goal. For example, in a sliding-block puzzle, the goal could be to get a particular block to a particular location. We use the terms "puzzle" and "one-player game" interchangeably. For puzzles, the generic forced-win decision question—"does player X have a forced win?"—becomes "is this puzzle solvable?"

There are many existing complexity results for puzzles. Often puzzles are NP-complete; Sudoku [175] and Peg Solitaire [168] are popular examples. Unbounded games, such as Rush Hour [56] and Sokoban [33], can be PSPACE-complete.

A puzzle is a natural model of nondeterministic computation: for each move, there is a choice. If the choices are all correct, then the puzzle is solved (if it is solvable). This is how nondeterministic Turing machines compute as well: they "guess" the next transition, "trying" to find an accepting computation history.

Constraint Logic. The one-player version of constraint logic is called Nondeterministic Constraint Logic (NCL). The rules are simply that on his turn, the player may reverse any edge resulting in a legal configuration, and the decision question is whether a given edge may ever be reversed. For the bounded version, we allow each edge to reverse at most once.

The unbounded version of NCL was inspired by Flake and Baum's "Generalized Rush Hour Logic" (GRL) [56], and in fact GRL incorporates gadgets with the same properties as our AND and OR vertices. GRL also requires a crossover gadget, however, which NCL does not—in the approach we take for showing PSPACE-completeness, it is possible to cross signals using merely AND and OR vertices. GRL uses a different notion of signal (dual-rail logic), for which this approach is not possible. The inherent flavor of GRL is also quite different from that of NCL; at a conceptual level, GRL requires inverters, and so is not monotone. However, formally it is the case that NCL is merely GRL reformulated as a graph problem, without a crossover gadget, and Flake and Baum deserve the credit for the original PSPACE-completeness proof. We take a different, simpler approach for showing PSPACE-completeness, reducing from QBF. Flake and Baum explicitly build a space-bounded, reversible computer.

Due to the simplicity of NCL, and the abundance of puzzles with reversible moves, it is often straightforward to find reductions showing various puzzles PSPACE-hard. This is the largest class of reductions presented in Part II.

5.1   Bounded Games

Bounded one-player games are puzzles in which there is a polynomial bound (typically linear) on the number of moves that can be made. Usually there is some resource that is used up. For example, in a game of Sudoku, the grid eventually fills up with numbers, and then either the puzzle is solved or it is not. In Peg Solitaire, each jump removes one peg, until eventually no more jumps can be made.

The nondeterminism of these games, plus the polynomial bound, means that they are in NP—a nondeterministically guessed solution can be checked for validity in polynomial time.

Bounded Nondeterministic Constraint Logic (Bounded NCL) is formally defined as follows:

BOUNDED NONDETERMINISTIC CONSTRAINT LOGIC (BOUNDED NCL)

Instance: Constraint graph G, edge e in G.

Question: Is there a sequence of moves on G that eventually reverses e, such that each edge is reversed at most once?

A Bounded NCL graph abstracts the essence of a bounded puzzle; it also serves as a concise model of polynomial-time-bounded nondeterministic computation.

Bounded NCL reverts essentially to Satisfiability (SAT), which is NP-complete. However, the standard constraint-logic crossover gadget still works, which makes reductions from Bounded NCL to bounded puzzles much more straightforward than direct reductions from SAT. Planar SAT is also NP-complete [115], but that result is not generally useful for puzzle reductions. For Planar SAT, the graph corresponding to the formula is a planar bipartite graph, with vertex nodes and clause nodes, plus a loop connecting the vertex nodes. The clause nodes are not connected, however. In contrast, a Bounded NCL graph corresponding to a Boolean formula feeds all the AND outputs into one final AND; reversing that final AND's output edge is possible just when the formula is satisfiable. Typically, this is a critical structure for puzzle reductions, because the victory condition is usually a local property (such as moving a black to a particular place) rather than a distributed property of the entire configuration. Thus, puzzle reductions typically require the construction of a crossover gadget, even though Planar SAT is NP-complete, and the planarity result for NCL is thus stronger in a sense than that for SAT.

5.1.1   NP-completeness

We reduce 3SAT (Section B.5.2) to Bounded NCL to show NP-hardness. Given an instance of 3SAT (a Boolean formula F in 3CNF), we construct a constraint graph G with an edge e that can be eventually reversed just when F is satisfiable.

Constructing G is straightforward. For each variable in F we have one CHOICE (red-red-red) vertex; for each OR in F we have an OR vertex; for each AND in F we have an AND vertex. At each CHOICE, one output corresponds to the asserted form of the corresponding variable; the other corresponds to the negated form. The CHOICE outputs are connected to the OR inputs, using FANOUTs as needed. The outputs of the ORs are connected to the inputs of the ANDs. Finally, there will be one AND whose output corresponds to the truth of F. A sample graph representing a formula is shown in Figure 5.1.

Theorem 5.1. *Bounded NCL is NP-complete.*

Proof: Given an instance of 3SAT (a Boolean formula F in 3CNF), we construct graph G as described above.

It is clear that if F is satisfiable, the CHOICE vertex edges may be reversed in correspondence with a satisfying assignment, such that the output edge may eventually be reversed. Similarly, if the output edge

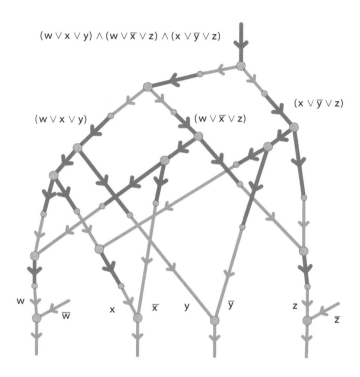

Figure 5.1. A constraint graph corresponding to the formula $(w \vee x \vee y) \wedge (w \vee \overline{x} \vee z)$ $\wedge (x \vee \overline{y} \vee z)$. Edges corresponding to literals, clauses, and the entire formula are labeled.

may be reversed, then a satisfying assignment may be read directly off the CHOICE vertex outputs.

Bounded NCL is also clearly in NP. Since each edge can reverse at most once, there can only be polynomially many moves in any valid solution; therefore, we can guess a solution and verify it in polynomial time. □

5.1.2 Restricted Problem

For reductions from Bounded NCL to other games, we need to strengthen Theorem 5.1 to apply to planar graphs that use a restricted set of vertex types. As mentioned above, the result that Planar SAT is NP-complete is not useful to us for either constraint graphs or actual games. But it is possible to build a crossover gadget within NCL.

Theorem 5.2. *Bounded NCL is NP-complete, even for planar graphs using only the vertex types shown in Figure 5.2.*

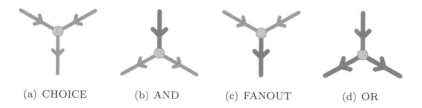

(a) CHOICE   (b) AND   (c) FANOUT   (d) OR

Figure 5.2. Basis vertices for Bounded NCL.

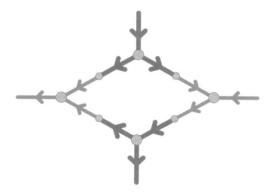

Figure 5.3. A half-crossover gadget that uses CHOICE vertices. Compare to
Figure 2.2(b).

Proof: In addition to the vertices shown in Figure 5.2, the reduction above
uses only red-blue vertices and loose edges. We use the standard conversion
techniques of Section 2.3 to eliminate these.

For planarity, the standard crossover gadget shown in Section 2.2 (on
page 20) is sufficient. Rather than give an explicit proof of correctness here
and then a more general proof for the unbounded case below, we merely
point out that the proof for the unbounded case, of Lemma 5.10, also applies to the bounded case; in the described sequences, no edge need ever
reverse more than once. There is one subtlety: the required initial edge orientations of the half-crossover gadget produce an AND-vertex subtype that
does not appear in Figure 5.2. However, the half-crossover can actually be
simplified by using CHOICE vertices, as shown in Figure 5.3. Here we have
simply used the CHOICE-vertex substitution from Section 2.3. This version
of the half-crossover (when we further translate the red-blue vertices) does
only use the permitted vertices.   □

5.1.3   Variant Problems

It will turn out to be useful to reduce from graphs that have the property that only a single edge can initially reverse. For this, we will have to explicitly add loose edges and red-blue vertices to the gadget set to reduce from. (As explained in Section 2.3, this generally costs nothing in actual game reductions.) Then, we simply take a single loose edge and split it enough times to reach all the free CHOICE inputs in the reduction.

Theorem 5.3. *Theorem 5.2 remains true when the input graph additionally uses red-blue vertices, and a single loose edge, which is the only edge that may initially reverse.*

Proof: As above.   □

A related problem is Constraint Graph Satisfiability:

CONSTRAINT GRAPH SATISFIABILITY

Instance: Planar constraint graph G using only AND and OR vertices.

Question: Does G have a configuration that satisfies all the constraints?

Properly, this problem is not a constraint-logic game, because the moves (assignments of edge orientations) are not reversals from one legal configuration to another. But it is similar in spirit and can prove useful for reductions.

Theorem 5.4. *Constraint Graph Satisfiability is NP-complete.*

Proof: We use essentially the same reduction as in Theorem 5.1; we change the graph slightly by adding a constrained edge terminator (Section 2.3) to the output edge. Then, if the formula is satisfiable, there is clearly a legal graph configuration, because there is a sequence of moves directing the output edge outward. If the formula is not satisfiable, then there is no legal graph configuration: the constrained edge terminator ensures that the output edge is directed outward from its final AND, which is only possible if the CHOICE outputs encode a satisfying assignment.

The standard conversion techniques of Section 2.3 make the graph planar, and convert the CHOICE vertices in the reduction to (unoriented) ANDs and ORs. Again, the problem is clearly in NP: we can check consistency in polynomial time.   □

Note that for Constraint Graph Satisfiability, unlike proper Bounded NCL, only two types of vertices are needed.

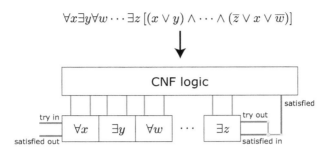

Figure 5.4. Schematic of the reduction from Quantified Boolean Formulas to NCL.

5.2 Unbounded Games

Unbounded one-player games are puzzles in which there is no restriction on the number of moves that can be made. Typically the moves are reversible. For example, in a sliding-block puzzle, the pieces may be slid around in the box indefinitely, and a block once slid can always be immediately slid back to its previous position.

Since there is no polynomial bound on the number of moves required to solve the puzzle, it is no longer possible to verify a proposed solution in polynomial time—the solution could have exponentially many moves. Indeed, unbounded puzzles are often PSPACE-complete. It is clear that such puzzles can be solved in nondeterministic polynomial space (NPSPACE), by nondeterministically guessing a satisfying sequence of moves; the only state required is the current configuration and the current move. But *Savitch's Theorem* [146] says that PSPACE = NPSPACE, so these puzzles can also be solved using deterministic polynomial space.

Nondeterministic Constraint Logic (NCL) is formally defined as follows:

NONDETERMINISTIC CONSTRAINT LOGIC (NCL)

Instance: Constraint graph G_0, edge e in G.

Question: Is there a sequence of moves on G that eventually reverses e?

5.2.1 PSPACE-completeness

We show that NCL is PSPACE-hard by giving a reduction from Quantified Boolean Formulas (QBF; see Section B.5.3). First we give an overview of the reduction and the necessary gadgets; then we analyze the gadgets' properties. The reduction is illustrated schematically in Figure 5.4. We translate a given quantified Boolean formula F in CNF into a constraint graph, so that a particular edge in the graph may be reversed if and only if F is true. The reduction is similar to the one for Deterministic Constraint Logic, in Chapter 4, but a bit simpler.

One way to determine the truth of a quantified Boolean formula is as follows: Consider the initial quantifier in the formula. Assign its variable first to false and then to true, and for each assignment, recursively ask whether the remaining formula is true under that assignment. For an existential quantifier, return true if either assignment succeeds; for a universal quantifier, return true only if both assignments succeed. For the base case, all variables are assigned, and we only need to test whether the CNF formula is true under the current assignment.

This is essentially the approach used in the reduction. We define *quantifier gadgets*, which are connected together into a string, one per quantifier in the formula, as in Figure 5.5(a). The rightmost edges of one quantifier are identified with the leftmost edges of the next. (This is different from the corresponding reduction in Chapter 4, where the edges are distinct and join at a blue-blue vertex.) Each quantifier gadget outputs a pair of edges corresponding to a variable assignment. These edges feed into the *CNF network*, which corresponds to the unquantified formula. The output from the CNF network connects to the rightmost quantifier gadget; the output of the overall graph is the **satisfied out** edge from the leftmost quantifier gadget.

Quantifier Gadgets. When a quantifier gadget is *activated*, all quantifier gadgets to its left have fixed particular variable assignments, and only this quantifier gadget and those to the right are free to change their variable assignments. The activated quantifier gadget can declare itself *satisfied* if and only if the Boolean formula read from here to the right is true given the variable assignments on the left.

A quantifier gadget is activated by directing its **try in** edge inward. Its **try out** edge is enabled to be directed outward only if **try in** is directed inward, and its variable state is locked. A quantifier gadget may nondeterministically "choose" a variable assignment, and recursively "try" the rest of the formula under that assignment and those that are locked by quantifiers to its left. The variable assignment is represented by two output edges (x and $\bar{x}$), only one of which may be directed outward. For **satisfied out** to be directed outward, indicating that the formula from this quantifier on is currently satisfied, **satisfied in** must be directed inward.

We construct both existential- and universal-quantifier gadgets, described below, satisfying the above requirements.

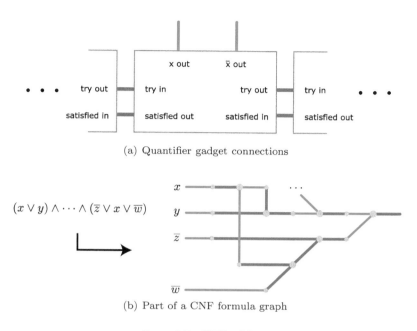

(a) Quantifier gadget connections

$(x \lor y) \land \cdots \land (\overline{z} \lor x \lor \overline{w})$

(b) Part of a CNF formula graph

Figure 5.5.   QBF wiring.

Lemma 5.5. *A quantifier gadget's* **satisfied in** *edge may not be directed inward unless its* **try out** *edge is directed outward.*

Proof: By induction. The condition is explicitly satisfied in the construction for the rightmost quantifier gadget, and each quantifier gadget requires **try in** to be directed inward before **try out** is directed outward, and requires **satisfied in** to be directed inward before **satisfied out** is directed outward.   □

CNF Formula. In order to evaluate the formula for a particular variable assignment, we construct an AND/OR subgraph corresponding to the unquantified part of the formula, fed inputs from the variable gadgets, and feeding into the **satisfied in** edge of the rightmost quantifier gadget, as in Figure 5.4. The **satisfied in** edge of the rightmost quantifier gadget is further protected by an AND vertex, so it may be directed inward only if **try out** is directed outward and the formula is currently satisfied.

Because the formula is in conjunctive normal form, and we have edges representing both literal forms of each variable (true and false), we do not need an inverter for this construction. Part of such a graph is shown in Figure 5.5(b). (Also see Figure 5.1.)

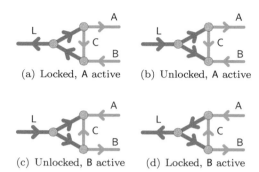

(a) Locked, **A** active   (b) Unlocked, **A** active

(c) Unlocked, **B** active   (d) Locked, **B** active

Figure 5.6. Latch gadget, transitioning from state **A** to state **B**.

Lemma 5.6. *The* **satisfied out** *edge of a CNF subgraph may be directed out-ward if and only if its corresponding formula is satisfied by the variable assignments on its input edge orientations.*

Proof: Definition of AND and OR vertices, and the CNF construction de-scribed.   □

Latch Gadget. Internally, the quantifier gadgets use *latch gadgets*, shown in Figure 5.6. This subgraph effectively stores a bit of information, whose state can be locked or unlocked. With edge **L** directed left, one of the other two OR edges must be directed inward, preventing its output red edge from pointing out. The orientation of edge **C** is fixed in this state. When **L** is directed inward, the other OR edges may be directed outward, and the red edges are free to reverse. Then when the latch is locked again, by directing **L** left, the state has been switched.

Existential Quantifier. An existential-quantifier gadget (Figure 5.7(a)) uses a latch subgraph to represent its variable, and beyond this latch has the minimum structure needed to meet the definition of a quantifier gadget. If the formula is true under some assignment of an existentially quantified variable, then its quantifier gadget may lock the latch in the corresponding state, enabling **try out** to activate, and recursively receive the **satisfied in** sig-nal. Receiving the **satisfied in** signal simultaneously passes on the **satisfied out** signal to the quantifier on the left.

   Here we exploit the nondeterminism in the model to choose the correct variable assignment.

Universal Quantifier. A universal-quantifier gadget is more complicated (Fig-ure 5.7(b)). It may only direct **satisfied out** leftward if the formula is true

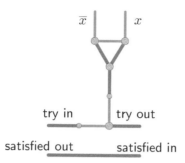

(a) Existential quantifier

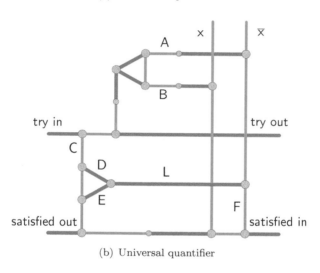

(b) Universal quantifier

Figure 5.7. Quantifier gadgets.

under both variable assignments. Again we use a latch for the variable state; this time we split the variable outputs, so they can be used internally. In addition, we use a latch internally, as a memory bit to record that one variable assignment has been successfully tried. With this bit set, if the other assignment is then successfully tried, **satisfied out** is allowed to point out.

Lemma 5.7. *A universal-quantifier gadget may direct its **satisfied out** edge outward if and only if at one time its **satisfied in** edge is directed inward while its variable state is locked in the false ($\overline{x}$) assignment, and at a later time the **satisfied in** edge is directed inward while its variable state is locked in the true (x) assignment, with **try in** directed inward throughout.*

Proof: First we argue that, with **try in** directed outward, edge **E** must point right. The **try out** edge must be directed inward in this case, so by Lemma 5.5, **satisfied in** must be directed outward. As a consequence, **F** and thus **L** must point right. On the other hand, **C** must point up and thus **D** must point left. Therefore, **E** is forced to point right in order to satisfy its OR vertex.

Suppose that **try in** is directed inward, the variable is locked in the false state (edge **A** points right), and **satisfied in** is directed inward. These conditions allow the internal latch to be unlocked, by directing edge **L** left. With the latch unlocked, edge **E** is free to point left. The latch may then lock again, leaving **E** pointing left (because **C** may now point down, allowing **D** to point right). Now, the entire edge reversal sequence that occurred between directing **try out** outward and unlocking the internal latch may be reversed. After **try out** has deactivated, the variable may be unlocked, and change state. Then, suppose that **satisfied in** activates with the variable locked in the true state (edge **B** points right). This condition, along with edge **E** pointing left, is both necessary and sufficient to direct **satisfied out** outward.   □

The behavior of both types of quantifiers is captured with the following property:

Lemma 5.8. *A quantifier gadget may direct its* **satisfied out** *edge out if and only if its* **try in** *edge is directed in, and the formula read from the corresponding quantifier to the right is true given the variable assignments that are fixed by the quantifier gadgets to the left.*

Proof: By induction. By Lemmas 5.5 and 5.7, if a quantifier gadget's **satisfied in** edge is directed inward and the above condition is inductively assumed, then its **satisfied out** edge may be directed outward only if the condition is true for this quantifier gadget as well. For the rightmost quantifier gadget, the precondition is explicitly satisfied by Lemma 5.6 and the construction in Figure 5.4.   □

Theorem 5.9. *NCL is PSPACE-complete.*

Proof: The graph is easily seen to have a legal configuration with the quantifier **try in** edges all directed leftward; this is the input graph G. The leftmost quantifier's **try in** edge may freely be directed rightward to activate the quantifier. By Lemma 5.8, the **satisfied out** edge of the leftmost quantifier gadget may be directed leftward if and only if F is true. Therefore, deciding whether that edge may reverse also decides the QBF problem, so NCL is PSPACE-hard.

NCL is in PSPACE because the state of the constraint graph can be described in a linear number of bits, specifying the direction of each edge,

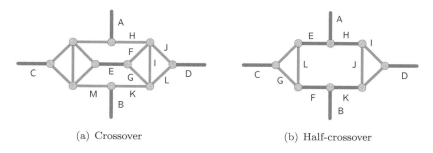

(a) Crossover   (b) Half-crossover

Figure 5.8. Planar crossover gadgets.

and the list of possible moves from any state can be computed in polynomial time. Thus we can nondeterministically traverse the state space, at each step nondeterministically choosing a move to make, and maintaining the current state but not the previously visited states. Savitch's Theorem [146] says that this NPSPACE algorithm can be converted into a PSPACE algorithm.   □

5.2.2 Restricted Problem

For reductions from NCL to other games, we need to strengthen Theorem 5.9 to apply to planar graphs that use a restricted set of vertex types. We begin by showing how to cross edges. This is the basic constraint-logic crossover proof, of which all the others are simple variants.

Figure 5.8(a) (which is the same as Figure 2.2(a)) illustrates the reduction. In addition to AND and OR vertices, this subgraph contains red-red-red-red vertices; these need any two edges to be directed inward to satisfy the inflow constraint of 2.

Lemma 5.10. *In a crossover subgraph, each of the edges A and B may face outward if and only if the other faces inward, and each of the edges C and D may face outward if and only if the other faces inward.*

Proof: We show that edge B can face down if and only if A does, and D can face right if and only if C does. Then by symmetry, the reverse relationships also hold.

Suppose A faces up, and assume without loss of generality that E faces left. Then so do F, G, and H. Because H and F face left, I faces up. Because G and I face up, K faces right, so B must face up. Next, suppose D faces right, and assume without loss of generality that I faces down. Then J and F must face right, and therefore so must E. An identical argument shows that if E faces right, then so does C.

Suppose A faces down. Then H may face right, I may face down, and K may face left (because E and D may not face away from each other). Symmetrically, M may face right; therefore B may face down. Next, suppose D faces left, and assume without loss of generality that B faces up. Then J and L may face left, and K may face right. Therefore G and I may face up. Because I and J may face up, F may face left; therefore, E may face left. An identical argument shows that C may also face left.   □

Next, we show how to represent the degree-4 vertices in Figure 5.8(a) with equivalent subgraphs using only ANDs and ORs. The necessary subgraph is shown in Figure 5.8(b). (This is the same subgraph as Deterministic Constraint Logic crossover gadget (Figure 4.7), but the different rules mean that a more complex crossover is necessary for NCL.) Note that red-blue vertices are necessary when substituting this subgraph into the previous one; the terminal edges in Figure 5.8(b) are blue, but it replaces red-red-red-red vertices. We must be careful to keep the graph planar when performing the red-blue reduction shown in Figure 2.4. But this is easy; we pair up edges A and D, and edges B and C.

Lemma 5.11. *In a half-crossover gadget, at least two of the edges A, B, C, and D must face inward; any two may face outward.*

Proof: Suppose that three edges face outward. Without loss of generality, assume that they include A and C. Then E and F must face left. This forces H to face left and I and J to face up; then D must face left and K must face right. But then B must face up, contradicting the assumption.

Next we must show that any two edges may face outward. We already showed how to face A and C outward. A and B may face outward if C and D face inward: we may face G and L down, F and K right, I and J up, and H and E left, satisfying all vertex constraints. Also, C and D may face outward if A and B face inward; the obvious orientations satisfy all the constraints. By symmetry, all of the other cases are also possible.   □

The crossover subgraph has blue free edges; what if we need to cross red edges, or a red and a blue edge? For crossing red edges, we may attach red-blue conversion subgraphs to the crossover subgraph in two pairs, as we did for the half-crossover. We may avoid having to cross a red edge and a blue edge, as follows: replace one of the blue edges with a blue-red-blue edge sequence, using a dual red-blue conversion subgraph. Then the original blue edge may be effectively crossed by crossing two red edges instead.

Theorem 5.12. *NCL is PSPACE-complete, even for planar graphs using only the vertex types shown in Figure 5.9.*

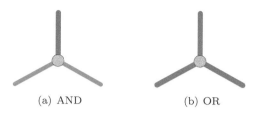

(a) AND   (b) OR

Figure 5.9. Basis vertices for NCL.

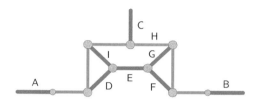

Figure 5.10. OR vertex made with protected-OR vertices.

Proof: Lemmas 5.10 and 5.11 show planarity. Any crossing edge pairs may be replaced by the above constructions; a crossing edge may be reversed if and only if a corresponding crossover edge (e.g., **A** or **C**) may be reversed. We must also specify configurations in the replacement graph corresponding to source configurations, but this is easy: pick any legal configuration of the crossover subgraphs with matching crossover edges.

To show that AND and OR vertices are sufficient, we simply note that the standard conversion techniques of Section 2.3 suffice to remove the red-blue vertices and the free edges **try in** and **satisfied out** in Figure 5.4. □

5.2.3 Protected-OR Graphs

So far we have shown that NCL is PSPACE-complete for planar constraint graphs using only AND and OR vertices. It is useful to make the conditions required for PSPACE-completeness still weaker; this will make the puzzle reductions in Chapter 9 simpler.

We call an OR vertex *protected* if there are two of its edges that, due to global constraints, cannot simultaneously be directed inward. Intuitively, graphs with only protected ORs are easier to reduce to another problem domain, since the corresponding OR gadgets need not function correctly in all the cases that a true OR must. We can simulate an OR vertex with a subgraph all of whose OR vertices are protected, as shown in Figure 5.10.

Lemma 5.13. *Edges A, B, and C in Figure 5.10 satisfy the same constraints as an OR vertex; all ORs in this subgraph are protected.*

Proof: Suppose that edges A and B are directed outward. Then D and F must be directed away from E. Assume without loss of generality that E points left. Then so must G; this forces H right and C down, as required. Then, by pointing A, D, and E right, we can direct G right, H left, and C up. Symmetrically, we can direct A and C out, and B in.

The two OR vertices shown in the subgraph are protected: edges I and D cannot both be directed inward, due to the red edge they both touch; similarly, G and F cannot both be directed inward. The red-blue conversion subgraph (Figure 2.4) we need for the two red-blue vertices also contains an OR vertex, but this is also protected.   □

Theorem 5.14. *Theorem 5.12 remains valid even when all of the OR vertices in the input graph are protected.*

Proof: Lemma 5.13. Any OR vertex may be replaced by the above construction; an OR edge may be reversed if and only if a corresponding subgraph edge (A, B, or C) may be reversed. We must also specify configurations in the replacement graph corresponding to source configurations: pick any legal configuration of the subgraphs with matching edges.   □

5.2.4   Configuration-to-Configuration Problem

Occasionally it will be desirable to reduce NCL to a problem in which the goal is to achieve some particular total state, rather than an isolated partial state. For example, in many sliding-block puzzles the goal is to move a particular piece to a particular place, but in others it is to reach some given complete configuration. One version of such a problem is the Warehouseman's Problem (Section 9.4).

For such problems, we show that an additional variant of NCL is hard.

Theorem 5.15. *Theorem 5.14 remains valid when the decision question is whether there is a sequence of moves on G that reaches a new state G',
rather than reverses an edge e.*

Proof: Instead of terminating **satisfied out** in Figure 5.4 with a free-edge terminator, attach a latch gadget (Figure 5.6), with free-edge terminators on its loose red edges. Then, it is possible to reach the initial state modified so that only the latch state is reversed just when it is possible to reverse **satisfied out**: first reverse **satisfied out** (by solving the QBF problem), unlocking the latch; then reverse the latch state; then undo all the moves that reversed **satisfied out**.   □

Two-Player Games

In this chapter we present the definitions and complexity proofs for Two-Player Constraint Logic, both the bounded and unbounded varieties.

Complexity Background. With two-player games, we are finally in territory familiar to classical game theory and combinatorial game theory. Two-player, perfect-information games are also the richest source of existing hardness results for games. In a two-player game, players alternate making moves, each trying to achieve some particular objective. The standard decision question is "does player X have a forced win from this position?"

The earliest hardness results for two-player games were PSPACE-completeness results for bounded games, beginning with Generalized Hex [53], and continuing with several two-player versions of known NP-complete problems [147]. Later, when the notion of alternating computation was developed [22], there were tools to show unbounded two-player games EXPTIME-complete. Chess [58], Go [143], and Checkers [145] then fell in quick succession to these techniques (though all three reductions are very complicated).

The connection between two-player games and computation is quite manifest. Just as adding the concept of nondeterminism to deterministic computation creates a new useful model of computation, adding an "extra degree" of nondeterminism leads to the concept of *alternating nondeterminism*, or *alternation* [22], discussed in Appendix B. Indeed, up to this point it is clear that adding an extra degree of nondeterminism is like adding an extra player in a game, and seems to raise the computational complexity of the game, or the computational power of the model of computation. Unfortunately this process does not generalize in the obvious way: simply adding extra players beyond two does not alter the situation in any fundamental way, from a computational-complexity standpoint. But later we will find other ways to add computational power to games.

Alternation raises the complexity of bounded games from the one-player complexity of NP-complete to PSPACE-complete, and of unbounded games from the one-player complexity of PSPACE-complete to EXPTIME-complete [157]. (These correspondences are clearer when expressed in terms of alternating complexity classes: AP = PSPACE; APSPACE = EXP-TIME.) Since it is not known whether P = NP or even PSPACE, with two-player games we finally reach games that are provably intractable: P $\neq$ EXPTIME [83]. In each case there is a natural game played on a Boolean formula that is complete for the appropriate class. For bounded games the game is equivalent to the Quantified Boolean Formulas problem: the "existential" and "universal" players take turns choosing assignments of successive variables. The unbounded games are similar, except that variable assignments can be changed back and forth multiple times.

Constraint Logic. For the two-player version of constraint logic—Two-Player Constraint Logic (2CL)—we create different moves for the two players, Black and White, by labeling each constraint graph edge as either Black or White. (This is independent of the red/blue coloration, which is simply a shorthand for edge weight.) Black (White) is allowed to reverse only Black (White) edges. As before, a move must reverse exactly one edge and result in a valid configuration. Each player has a target edge he is trying to reverse.

6.1   Bounded Games

Bounded two-player games are games in which there is a polynomial bound (typically linear) on the number of moves that can be made. As with bounded puzzles, usually there is some resource that is used up. In Hex, for example, each move fills a space on the board, and when all the spaces are full, the game must be over. Similarly, in Amazons, on each move an amazon must remove one of the spaces from the board. In Konane, each move removes at least one stone. There are many other bounded two-player games. In these games, when the resource is exhausted, the game cannot continue.

Deciding such games can be no harder than PSPACE, because a Turing machine using space polynomial in the board size can perform a depth-first search of the entire game tree, determining the winner. In general these games are also PSPACE-hard. The canonical PSPACE-completegame is simply Quantified Boolean Formulas (QBF). The question "does there exist an x, such that for all y, there exists a z, such that ... formula F is true" is equivalent to the question of whether the first player can win the following formula game: "Players take turns assigning truth values to a sequence of variables. When they are finished, player one wins if formula F is true; otherwise, player two wins."

Bounded Two-Player Constraint Logic is the natural form of constraint logic that corresponds to this type of game. It is formally defined as follows:

BOUNDED TWO-PLAYER CONSTRAINT LOGIC (BOUNDED 2CL)

Instance: Constraint graph G, partition of the edges of G into sets B and W, and edges $e_B \in B$, $e_W \in W$.

Question: Does White have a forced win in the following game? Players White and Black alternately make moves on G. White (Black) may only reverse edges in W (B). Each edge may be reversed at most once. White (Black) wins (and the game ends) if he ever reverses e_W (e_B).

6.1.1   PSPACE-completeness

Many variants of the basic QBF problem are also PSPACE-complete [147]. It will be most convenient to reduce one of these variants to Bounded 2CL to show PSPACE-hardness, rather than reducing directly from the standard form of QBF:

G_{pos}(POS CNF)

Instance: Monotone CNF formula A (that is, a CNF formula in which there are no negated variables).

Question: Does Player I have a forced win in the following game? Players I and II alternate choosing some variable of A that has not yet been chosen. The game ends after all variables of A have been chosen. Player I wins if and only if A is true when all variables chosen by Player I are set to true and all variables chosen by II are set to false.

The reduction from G_{pos}(POS CNF) to Bounded 2CL is similar to the reduction from SAT to Bounded NCL in Section 5.1. There, the single

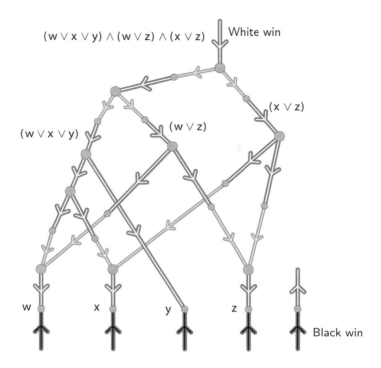

Figure 6.1.   A constraint graph corresponding to the G_{pos}(POS CNF) formula game $(w \lor x \lor y) \land (w \lor z) \land (x \lor z)$. Edges corresponding to variables, clauses, and the entire formula are labeled.

player is allowed to choose a variable assignment via a set of CHOICE vertices. All we need do to adapt this reduction is replace the CHOICE vertices with VARIABLE vertices, such that if White plays first in a variable vertex the variable is true, and if Black plays first the variable is false. Then, we attach White's variable vertex outputs to the CNF formula inputs as before; Black's variable outputs are unused. The CNF formula consists entirely of White edges. Black is given enough extra edges to ensure that he will not run out of moves before White. White's target edge is the formula output, and Black's is an arbitrary edge that is arranged to never be reversible. A sample game graph corresponding to a formula game is shown in Figure 6.1; compare to Figure 5.1. (The extra Black edges are not shown.) Note that the edges are shown filled with the color that controls them.

The game breaks down into two phases. In the first phase, players alternate playing in variable vertices, until all have been played in. Then,

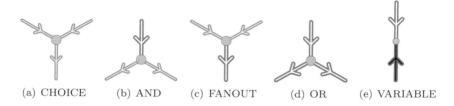

(a) CHOICE   (b) AND   (c) FANOUT   (d) OR   (e) VARIABLE

Figure 6.2. Basis vertices for Bounded 2CL.

White will win if he has chosen a set of variables satisfying the formula. Since the formula is monotone, it is exactly the variables assigned to be true, that is, the ones White chose, that determine whether the formula is satisfied. Black's play is irrelevant after this.

Theorem 6.1. *Bounded 2CL is PSPACE-complete.*

Proof: Reduction from $G_{\text{pos}}(\text{POS CNF})$, as described above. If Player I can win the formula game, then White can win the corresponding Bounded 2CL game, by playing the formula game on the edges, and then reversing the necessary remaining edges to reach the target edge. If Player I cannot win the formula game, then White cannot play so as to make a set of variables true that will satisfy the formula, and thus he cannot reverse the target edge. Neither player can benefit from playing outside the variable vertices until all variables have been selected, because this can only allow the opponent to select an extra variable.

This shows that Bounded 2CL is PSPACE-hard. It is also clearly in PSPACE: since the game can only last as many moves as there are edges, a simple depth-first traversal of the game tree suffices to determine the winner from any position.   □

6.1.2 Restricted Problem

For reductions from Bounded 2CL to other games, we need to strengthen Theorem 6.1 to apply to planar graphs that use a restricted set of vertex types. Indeed, the above reduction is almost trivial, and the true benefit of using Bounded 2CL for game reductions, rather than simply using one of the many QBF variants, is that when reducing from Bounded 2CL one does not have to build a crossover gadget. The bounded two-player games addressed in Part II have relatively straightforward reductions for this reason. The complexity of Amazons remained open for several years, despite some effort by the game complexity community.

Theorem 6.2. *Bounded 2CL is PSPACE-complete, even for planar graphs using only the vertex types shown in Figure 6.2.*

Proof: For planarity, the crossover gadget presented in Section 5.2.2 is again sufficient. Note that no Black edges ever cross; therefore, all crossovers are monochrome, and essentially one-player crossovers.

For the vertex types, apart from VARIABLE vertices, the reduction in Section 6.1.1 uses the same types of vertex as that in Section 5.1.1, and the same vertex substitutions work. (Note that while the CHOICE vertex does not occur directly in the present reduction, it does occur inside the crossover gadget, as described in Section 5.1.1.)   □

6.2   Unbounded Games

Unbounded two-player games are games in which there is no restriction on the number of moves that can be made. Typically (but not always) the moves are reversible. Examples include the classic games Chess, Checkers, and Go. Some moves in each of these games are not reversible: pawn movement, castling, capturing, and promoting, in Chess; normal piece movement, capturing, and kinging in Checkers; and, actually, every move in Go. Go is an interesting case, because at first sight it appears to be a bounded game: every move places a stone, and when the board is full the game is over. However, capturing removes stones from the board, and reopens the spaces they occupied. Each of these games is EXPTIME-complete [58, 143, 145].[1]

Indeed, there are "proofs" that Go is PSPACE-complete, including one by Papadimitriou [132, pages 462–469]. Papadimitriou does not make the mistake of thinking that Go is a bounded game, however; instead, he considers a modified version that is bounded. In fact, Go's peculiarities, combined with the extreme simplicity of the rules, make it worthy of extra study from a complexity standpoint.

The removal of a polynomial bound on the length of the game means that it is no longer possible to perform a complete search of the game tree using polynomial space, so the PSPACE upper bound no longer applies. In general, two-player games of perfect information are EXPTIME-complete [157].

Two-Player Constraint Logic (2CL) is the form of constraint logic that corresponds to this type of game. It is formally defined as follows:

---

[1] For Go, the result is only for Japanese rules. See Section 6.3.

TWO-PLAYER CONSTRAINT LOGIC (2CL)

Instance: AND/OR constraint graph G, partition of the edges of G into sets B and W, and edges $e_B \in B$, $e_W \in W$.

Question: Does White have a forced win in the following game? Players White and Black alternately make moves on G. White (Black) may only reverse edges in W (B). White (Black) wins if he ever reverses e_W (e_B).

6.2.1 EXPTIME-completeness

To show that 2CL is EXPTIME-hard, we reduce from one of the several Boolean formula games that were shown to be EXPTIME-complete by Stockmeyer and Chandra [157]:

G_6

Instance: CNF Boolean formula F in variables $X \cup Y$, $(X \cup Y)$ assignment α.

Question: Does Player I have a forced win in the following game? Players I and II take turns. Player I (II) moves by changing at most one variable in X (Y); passing is allowed. Player I wins if F ever becomes true.

Note that there is no provision for Player II to ever win; the most he can hope to accomplish is a draw, by preventing Player I from ever winning. But this will not matter to us, because the relevant decision question for 2CL is simply whether White can force a win.

Reduction. The essential elements of the reduction from G_6 to 2CL are shown in Figure 6.3. This figure shows a White variable gadget and associated circuitry; a Black variable gadget is identical except that the edge marked **variable** is then black instead of white. The left side of the gadget is omitted; it is the same as the right side. The state of the variable depends on whether the **variable** edge is directed left or right, enabling White to reverse either the **false** or the **true** edge (and thus lock the **variable** edge into place).

The basic idea of the reduction is the same as for Bounded 2CL: the players should play a formula game on the variables, and then if White can win the formula game, he can then reverse a sequence of edges leading into the formula, ending in his target edge. In this case, however, the reduction is not so straightforward, because the variables are not fixed once chosen;

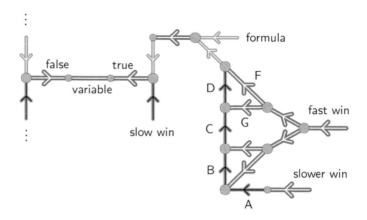

Figure 6.3. Reduction from G_6 to 2CL.

there is no natural mechanism in 2CL for transitioning from the variable-selection phase to the formula-satisfying phase. That is what the rest of the circuitry is for. (Also note that unlike the bounded case, the formula need not be monotone.)

White has the option, whenever he wishes, of *locking* any variable in its current state, without having to give up a turn, as follows. First, he moves on some **true** or **false** edge. This threatens to reach an edge F in four more moves, enabling White to reach a **fast win** pathway leading quickly to his target edge. Black's only way to prevent F from reversing is to first reverse D. But this would just enable White to immediately reverse G, reaching the target edge even sooner. First, Black must reverse A, then B, then C, and finally D; otherwise, White will be able to reverse one of the blue edges leading to the fast win. This sequence takes four moves. Therefore, Black must respond to White's **true** or **false** move with the corresponding A move, and then it is White's turn again.

The lengths of the pathways **slow win**, **slower win**, and **fast win** are detailed below in the proof. The pathways labeled **formula** feed into the formula vertices, culminating in White's target edge. It will be necessary to ensure that regardless of how the formula is satisfied, it always requires exactly the same number of edge reversals beginning with the formula input edges. The first step to achieving this is to note that the formula may be in CNF. Thus, every clause must have one variable satisfied, so it seems we are well on our way. However, there is a problem. Generally, a variable must pass through an arbitrary number of FANOUTs on its way into the clauses. This means that if it takes x reversals from a given variable gadget

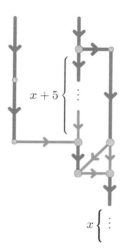

Figure 6.4. Path-length-equalizer gadget.

to a usage of that variable in a clause, it will take less than $2x$ reversals to reach two uses of the variable, and we cannot know in advance how many variables will be reused in different clauses. The solution to this problem is to use a path-length-equalizer gadget, shown in Figure 6.4. This gadget has the property that if it takes x reversals from some arbitrary starting point before entering the gadget, then it takes $x + 6$ reversals to reverse either of the topmost output edges, or $2x + 12$ reversals to reverse both of them. By using a chain of n such gadgets whenever a variable is used n times in the formula, we can ensure that it always takes the same number of moves to activate any variable instance in a clause, and thus that it always takes the same number of moves to activate the formula output.

Theorem 6.3. *2CL is EXPTIME-complete.*

Proof: Given an instance of G_6, we construct a corresponding constraint graph as described above.

Suppose White can win the formula game. Then, also suppose White plays to mimic the formula game, by reversing the corresponding **variable** edges up until reversing the last one that lets him win the formula game. Then Black must also play to mimic the formula game: his only other option is to reverse one of the edges **A** to **D**, but any of these lead to a White win.

Now, then, assume it is White's turn, and either he has already won the formula game, or he will win it with one more variable move. He proceeds to lock all the variables except possibly for the one remaining that he needs to change, one by one. As described above, Black has no choice but to respond to each such move. Finally, White changes the remaining variable, if needed. Then, on succeeding turns, he proceeds to activate the needed pathways through the formula and on to the target edge. With all the variables locked, Black cannot interfere. Instead, Black can try to activate one of the **slow win** pathways enabled during variable locking. However, the path lengths are arranged such that it will take Black one move longer to win on such a pathway than it will take White to win by satisfying the formula.

Suppose instead that White cannot win the formula game. He can accomplish nothing by playing on variables forever; eventually, he must lock one. Black must reply to each lock. If White locks all the variables, then Black will win, because he can follow a **slow win** pathway to victory, but White cannot reach his target edge at the end of the formula, and Black's **slow win** pathway is faster than White's **slower win** pathway. However, White may try to cheat by locking all the Black variables, and then continuing to change his own variables. But in this case Black can still win, because if White takes the time to change more than one variable after locking any variable, Black's **slow win** pathway will be faster than White's formula activation.

Thus, White can win the 2CL game if and only if he can win the corresponding G_6 game, and 2CL is EXPTIME-hard. Also, 2CL is easily seen to be in EXPTIME: the complete game tree has exponentially many positions, and thus can be searched in exponential time, labeling each position as a win, loss, or draw depending on the labels of its children. (A draw is possible when optimal play loops.) Therefore, 2CL is EXPTIME-complete. □

6.2.2 Restricted Problem

For reductions from 2CL to other games, we need to strengthen Theorem 6.3 to apply to planar graphs that use a restricted set of vertex types. In principle, this should enable much simpler reductions to actual games than the existing reductions from Boolean formula games. The existing Chess, Checkers, and Go hardness results are all quite complicated; we had hoped to re-derive some of them more simply using 2CL. We have not done so yet, however. Enforcing the necessary constraints in a two-player game gadget is much more difficult than in a one-player game.

The needed basis vertices are shown in Figure 6.5. Black's **slow win** pathway is implemented with alternating red and blue edges, joined with red-blue conversion gadgets; the OR vertices in these gadgets may be White. (Note that color symmetries in a target problem could mean that the Black AND gadget would be a color-swapped version of the White one, so there could be only five gadgets to build for a reduction.)

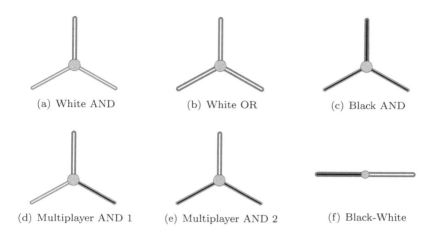

(a) White AND   (b) White OR   (c) Black AND

(d) Multiplayer AND 1   (e) Multiplayer AND 2   (f) Black-White

Figure 6.5. Basis vertices for 2CL.

Theorem 6.4. *2CL is EXPTIME-complete, even for planar graphs using only the vertex types shown in Figure 6.5.*

Proof: For planarity, the crossover gadget presented in Section 5.2.2 is again sufficient. Note that no Black edges ever cross; therefore, all crossovers are monochrome, and essentially one-player crossovers. It is possible that introduction of crossover gadgets can change some path lengths from variables through the formula; however, it is easy to pad all the variable pathways to correct for this, because it takes a fixed number of reversals to traverse a crossover gadget in each direction.

For the vertex types, again the standard conversion techniques of Section 2.3 are sufficient to remove the red-blue vertices and loose edges used in the reduction. All other vertex types used in the reduction appear in Figure 6.5.   □

6.3   No-Repeat Games

One interesting result deserves to be mentioned here. If the condition that no previous position may ever be recreated is added to two-player games, then the general complexity rises from EXPTIME-complete to EXPSPACE-complete [144]. The intuition is that it requires an exponential amount of space even to determine the legal moves from a position, because the game history leading up to a position could be exponentially long.

In fact, some of the EXPTIME-complete formula games in [157] automatically become EXPSPACE-complete when this modification is made, and as a result, no-repeat versions of Chess and Checkers are EXPSPACE-complete. However, the result does not apply to the game G_6, which was used to show 2CL EXPTIME-complete; nor does it apply to Go, which was shown EXPTIME-hard by a reduction from G_6. Go is actually played with this rule in many places; in Go it is called the *superko* rule. The complexity of Go with superko is an interesting problem that is still unresolved. Surprisingly, both the lower and the upper bounds of Robson's EXPTIME-completeness proof [143] break when the superko rule is added. Superko is not used in Japan; it is used in the US, China, and other places.

It is arguably a bit unnatural in a game for the set of legal moves to not be determinable from the current position. Of course, if the position is defined to include the game history then this is not a problem, but then the position grows over time, which is also against the spirit of generalized combinatorial games.

There is a tantalizing connection here to a kind of imperfect information, which is also connected to the idea of an additional player. A useful perspective is that in a two-person game, there is an "existential" player and a "universal" player. No-repeat games are almost like two-person games with an extra, "super-universal" player added. This player can remember, secretly, one game position. Then, if that position is ever recreated, whoever recreated it loses. In principle, this approach seems capable of resolving the above problems. All ordinary moves are legal, whether they are repeating or not, but in actual play repeating moves are losing because the super-universal player can nondeterministically guess in advance which position will be repeated. However, this idea seems difficult to formalize usefully; in particular, it is not clear how to formulate an appropriate decision question so that the super-universal player does not effectively team up with the universal player and against the existential one. But this seems an interesting path for further exploration.

Notwithstanding the above concerns, a no-repeat version of Two-Player Constraint Logic ought to be EXPSPACE-complete. A reduction from game G_3 from [157], for example, would do the trick, but we do not yet have one.

A Tempting Generalization. It is tempting to think of no-repeat games as fitting neatly into an ordered sequence: bounded, unbounded, no-repeat. (Indeed, this chapter is organized according to this conception.) For two-player games, the corresponding complexities are PSPACE-complete, EXPTIME-complete, EXPSPACE-complete, forming a natural sequence. Does this idea generalize to other numbers of players? Unfortunately not, at least for zero- and one-player games: if a one-player game can be solved at all, then clearly it can be solved without repeating any positions, and a similar observation applies to zero-player games.

Team Games

In this chapter we present the definitions and complexity proofs for Team Private Constraint Logic, both the bounded and unbounded varieties.

Complexity Background. It turns out that adding players beyond two to a game does not increase the complexity of the standard decision question, "does player X have a forced win?" We might as well assume that all the other players team up to beat X, in which case we effectively have a two-player game again. If we generalize the notion of the decision question somewhat, we do obtain new kinds of games. In a *team game*, there are still two "sides," but each side can have multiple players, and the decision question is whether *team X* has a forced win. A team wins if any of its players wins.

Team games with perfect information are still just two-player games in disguise, however, because again all the players on a team can cooperate and play as if they were a single player. However, when there is hidden information, then team games turn out to be different from two-player games.[1] (We could think of a team in this case as a player with a peculiar kind of mental limitation—on alternate turns he forgets some aspects of his

---

[1] Adding imperfect information to a two-player, unbounded game does create a new kind of game, intermediate in complexity between two-player perfect-information games and team games with imperfect information [136]; such games can be 2EXPTIME-complete (complete in doubly-exponential time) to decide. "Blindfold" and "hierarchical" games are introduced in [136], [133], and [134]. These games correspond to yet more complexity classes. These types of games have yet to be studied from a constraint-logic perspective.

situation, and remembers others.) Therefore, we will only consider team games of imperfect information, and we will sometimes simply refer to them as "team games."

Such games, as with two-player imperfect-information games, were first studied from a complexity standpoint by Peterson and Reif [133]. The general result is that bounded team games are NEXPTIME-complete, and unbounded games are undecidable. However, there are a few technical problems with the original undecidability result; in fact, the game claimed undecidable in [133], called TEAM-PEEK, is actually decidable. In Section 7.2 we show how to fix these problems, confirming that indeed team games of imperfect information are undecidable.

The fact that there are undecidable games using bounded space—when actually played, finite physical resources—at first seems counterintuitive and bizarre. There are only finitely many configurations in such a game. Eventually, the position must repeat. Yet, somehow the state of the game must effectively encode the contents of an unboundedly long Turing-machine tape! How can this be? These issues are discussed in Chapter 8.

Constraint Logic. The natural team, private-information version of constraint logic assigns to each player a set of edges he can control, and a set of edges whose orientation he can see. As always, each player has a target edge he must reverse to win. To enable a simpler reduction to the unbounded form of this game, we allow each player to reverse up to some given constant k edges on his turn, rather than just one, and leave the case of $k = 1$ as an open problem.

7.1 Bounded Games

Bounded team games of imperfect information include card games such as Bridge. Here we can consider one hand to be a game, with the goal being either to make the bid, or, if on defense, to set the other team. Focusing on a given hand also removes the random element from the game, making it potentially suitable for study within the present framework.

Bounded team games of private information are NEXPTIME-complete in general, by a reduction from Dependency Quantifier Boolean Formulas (DQBF) [133].

Bounded Team Private Constraint Logic (Bounded TPCL) starts from a configuration known to all players; the private information arises as a result of some moves not being visible to all players. (These attributes do not apply to Bridge directly, but some sort of reduction may be possible.)

BOUNDED TEAM PRIVATE CONSTRAINT LOGIC (BOUNDED TPCL)

Instance:

Constraint graph G; integer N; for $i \in \{1 \ldots N\}$: sets $E_i \subset V_i \subset G$, edges $e_i \in E_i$; partition of $\{1 \ldots N\}$ into nonempty sets W and B.

Question: Does White have a forced win in the following game? Players $1 \ldots N$ take turns in that order. Player i only sees the orientation of the edges in V_i, and moves by reversing an edge in E_i thathas not previously reversed; a move must be known legal based on V_i. White (Black) wins if Player $i \in W$ (B) ever reverses edge e_i.

7.1.1   NEXPTIME-completeness

We show that Bounded TPCL is NEXPTIME-complete by a reduction from Dependency Quantifier Boolean Formulas (DQBF), a generalization of QBF introduced by Peterson and Reif [133]. We state here a game version of DQBF, shown NEXPTIME-complete in [133]:

DEPENDENCY QUANTIFIER BOOLEAN FORMULA GAME (DQBF-GAME)

Instance: CNF Boolean formula F in variables $X_i \cup Y_i$ for $i \in \{1, 2\}$.

Question: Does White have a win in the following game? Player B (Black) chooses an assignment to $X_1 \cup X_2$, then Player W_1 (White) chooses an assignment to Y_1, then Player W_2 (White) chooses an assignment to Y_2. Player W_i only sees the assignments to X_i and Y_i. White (Black) wins if F is true (false) under the chosen variable assignments.

Theorem 7.1. *Bounded TPCL is NEXPTIME-complete.*

Proof: The reduction from DQBF-Game is shown in Figure 7.1. For convenience we let the White players be numbered 1 and 2, and the Black player 3. The game breaks down into two phases. In the first, Black chooses his variable assignments. In the second, the White players choose theirs, and try to satisfy the formula to reach their target edge. If they cannot, Black has time to reach his own target edge. We arrange it so that initially Black's only moves are to choose variable assignments. The White

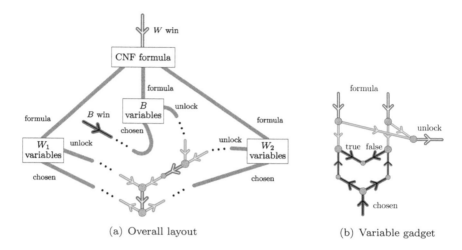

(a) Overall layout   (b) Variable gadget

Figure 7.1. Reduction from DQBF-GAME to Bounded TPCL.

players each have a pool of extra edges so that they can kill time while waiting for Black to choose variables.

The variable-selection gadget is shown in Figure 7.1(b). This represents a Black variable; the White variables are identical apart from edge colors. Each Black variable gadget represents a variable in X_i ($i = 1$ or 2); the edges in the gadget are then in V_i (visible to the corresponding White player), and the White edges in the gadget are in E_i (reversible by that player). Each White variable gadget represents a variable in Y_i; all of its edges are seen and controlled only by player i, except that the output **formula** edges are visible to both White players, as are the Black-variable **formula** edges. The formula AND/OR network, culminating in the White target edge, is seen and controlled by both White players.

Once a variable selection has been made, true or false, the **chosen** output edge for that variable may be activated. All the **chosen** edges for a given player are fed into a series of AND vertices, with a single output edge (**all-chosen**) that can be reversed just when all the player's variable selections have been made. For Black, this edge leads to a chain of edge reversals culminating in his target edge. For the White players, the two **all-chosen** edges feed into an AND. This AND output then splits by a series of FANOUTs to feed all of the **unlock** edges, allowing the variable selections to feed into the formula.

Suppose White can win the formula game. Then, Black must eventually select values in all his corresponding variable gadgets; each White player can then select his variable values based on the Black edges that are visible to him. When all White values have been chosen, privately from each other, they may then be unlocked (along with the Black variables) and fed into the formula, allowing White to win. If Black wins the formula game, then no variable selection by White will let them reverse their target edge, and Black will eventually reverse his target edge and win.

This shows NEXPTIME-hardness; we must still show that Bounded TPCL is in NEXPTIME. To do that, we show that we can verify, in exponential time, a nondeterministically guessed winning White strategy. The strategy will be a mapping from visible game histories to moves, for each White player. A visible game history is just a history where all moves of edges not visible to the player in question are simply recorded as "hidden." The number of possible histories is exponential in the graph size, therefore the strategy will be of exponential length as well.

To verify the strategy, we evaluate the complete game tree to test whether White always wins: for each Black move, consider all legal possibilities; for each White move, consider only the move that the corresponding strategy dictates. The game tree is also of exponential size, and thus it can be searched in exponential time. □

7.1.2 Restricted Problem

As usual, we strengthen Theorem 7.1 to apply to planar graphs that use a restricted set of vertex types.

Theorem 7.2. *Bounded TPCL is NEXPTIME-complete, even for planar graphs that use only AND and OR vertices.*

Proof: All crossings in the reduction involve only edges controlled by a single player or by both White players; we can replace these with crossover gadgets from Section 5.2.2 without changing the game. As usual, we can eliminate red-blue edges and loose edges with the techniques of Section 2.3. All remaining vertices are ANDs and ORs. (However, note that several different AND- and OR-subtypes are used in the reduction, which we do not enumerate.) □

7.2 Unbounded Games

In general, team games of private information are undecidable. This result was claimed by Peterson and Reif in 1979 [133]. However, as mentioned above, there are a few problems with the proof, which we address in Section 7.2. Strangely, the result also seems to be not very well known. Part of the problem may be that the authors seem to consider the result of secondary importance to the other results in [133]. Indeed, immediately after showing their particular game undecidable, the authors remark

> In order to eliminate, by restriction, the over-generality of MPA_k-TMs, we considered several interesting variants. [133, page 355]

The rest of the paper then considers those variants. From our perspective, however, the fact that there are undecidable space-bounded games is fundamental to the viewpoint that games are an interesting model of computation. It both shows that games are as powerful as general Turing machines, and highlights the essential difference from the Turing-machine foundation of theoretical computer science, namely that *a game computation is a manipulation of finite resources*. Thus, this seems to be a result of some significance.

Recall the discussion of Rengo Kriegspiel in Section 3.4.2. (See Appendix A for details.) The important difference from other undecidable problems, such as the Post Correspondence Problem (PCP), is that Rengo Kriegspiel is a game with bounded space; there are a fixed number of positions in any given game. Thus, the game can *actually* be played; PCP, though a puzzle of a sort, cannot be played in the real world without unbounded resources. This theme will be developed further in Chapter 8.

Team Private Constraint Logic is defined as follows. Note the addition of the parameter k relative to the bounded case. This is, admittedly, an extra generalization to make a reduction easier; nonetheless, it is a reasonable generalization, and all other constraint-logic games are naturally restricted versions of this game.

TEAM PRIVATE CONSTRAINT LOGIC (TPCL)

Instance: Constraint graph G; integer N; for $i \in \{1 \dots N\}$: sets $E_i \subset V_i \subset G$, edges $e_i \in E_i$; partition of $1 \dots N$ into nonempty sets W and B; integer k.

Question: Does White have a forced win in the following game? Players $1 \dots N$ take turns in that order. Player i only sees the orientation of the edges in V_i, and moves by reversing up to k edges in E_i; a move must be known legal based on V_i. White (Black) wins if Player $i \in W$ (B) ever reverses edge e_i.

Before showing this game undecidable, we discuss the earlier results of Peterson and Reif [133].

TEAM-PEEK.   A particular space-bounded game with alternating turns, TEAM-PEEK, is undecidable [133]. (TEAM-PEEK is a team version of

Stockmeyer and Chandra's EXPTIME-complete game PEEK [157].) There are two problems with this claim. First, there is a simple mistake in the definition, making the undecidability claim false. To see this, consider the following formal statement of TEAM-PEEK, which is equivalent to the more physical version described in [133]:

TEAM-PEEK

Instance: DNF Boolean formula F in variables S, integer N; for $i \in \{1 \ldots N\}$: sets $X_i \subset V_i \subset S$; partition of $1 \ldots N$ into nonempty sets W and B; S assignment α.

Question: Does White have a forced win in the following game? Players $1 \ldots N$ take turns in that order. Player i only knows the truth assignment to the variables in V_i, and moves by changing the truth assignment of any subset of the variables X_i. White (Black) wins if F is ever true after a move by Player $i \in W$ (B).

We show that TEAM-PEEK is decidable when White has two or more players and Black has one player (contrary to [133, Theorem 5]). Whatever the turn order, the White players will wind up playing in sequence. Now it is easy to tell whether White can win before Black's first move, so assume that they cannot. Then, either White can win immediately regardless of Black's move, which is also easy to determine, or they cannot. Suppose they cannot. Then, they cannot have a forced win at all, because whatever moves they make in sequence on any pair of turns, there is always some move Black could have just made that prevented a win.

The basic problem with the game definition is that allowing a player to change any or all of his variables in a single turn, instead of at most one variable as in PEEK, prevents threats and thus forcing moves. Thus, the standard machinery of building Turing-machine-acceptance reductions to formula games breaks down.

Round-Robin Play. The second problem has to do with the order in which players move. TEAM-PEEK's definition follows the natural form of game play in which players take turns in round robin. However, the problem developed in [133, page 355] for a reduction to TEAM-PEEK does not have this property:

> Given a TM M ... [t]he game ... will be based on having each of the $\exists$-players find a sequence of configurations of M which on an input that leads to acceptance. Hence, each $\exists$-player will

give to the ∀-player on request the next character of its sequence of configurations (secretly from the other). Each ∃-player does this secretly from the other ∃-player. The configuration will [be] in the form: $\#C_0\#C_1\#\ldots\#C_m\#$, where C_0 is the initial configuration of M on the input, and C_m is an accepting configuration of M.

The ∀-player will choose to verify the sequences in one of the following ways: ...

The ∀-player verifies the sequences by ensuring that the initial configurations match the input, that the final sequences are accepting, and that the transitions are valid. The existential team wins if a player generates a valid accepting history; the universal player wins if it detects an invalid history. The key is that the validity of the transitions can be checked with only a fixed amount of memory, by running one of the players ahead to the next $\#$ symbol, and stepping through both histories symbol by symbol.

It is implicit in the definition of the game that the universal player chooses, on each of his turns, which existential player is to play next, and the other existential player cannot know how many turns have elapsed before he gets to play again. For suppose instead that play does go round robin. Then we must assume that on the universal player's turn, he announces which existential player is to make a computation-history move this turn; the other one effectively passes on his turn. But then each existential player knows exactly where in the computation history the other one is, and whichever player is behind knows he cannot be checked for validity, and is at liberty to generate a bogus computation history. It is the very information about how many turns the other existential player has had that must be kept private for the game to work properly.

Reif has confirmed in a personal communication regarding round-robin play in TEAM-PEEK [138] that "it looks like therefore the players do not play round robin."[2] Indeed, the general definition of the game states, "Players need not take turns in a round-robin fashion. The rules will dictate whose turn is next ... A player may not know how many turns were taken by other players between its turns" [133, p. 349–350]. These notions deviate from (and hence exclude) the intuitive notion of a game, where players take turns in order and are aware of what happens between their turns. Therefore, we work to strengthen the approach to apply to this natural form of game.

---

[2]Both the mistake in definition described above and the turn-order problem also apply to all of the TEAM-PEEK variants defined in [134].

7.2.1 Undecidability.

To solve the above problems, we introduce a somewhat more elaborate computation game, in which the players take successive turns, and which we show to be undecidable. We reduce this game to a formula game, and the formula game to TPCL.

The new computation game will be similar to the above computation game, but each existential player will be required to produce successive symbols from two identical, independent computation histories, A and B; on each turn, the universal player will select which history each player should produce a symbol from, privately from the other player. Then, for any game history experienced by each existential player, it is always possible that his symbols are being checked for validity against the other player's, because one of the other existential player's histories could always be retarded by one configuration (or the history could be checked against the input). The fact that the other player has produced the same number of symbols as the current player does not give him any useful information, because he does not know the relative advancement of the other player's two histories.

This game is formalized as follows:

TEAM COMPUTATION GAME

Instance: Finite set of $\exists$-*options* O, Turing machine S with fixed tape length k, and with tape symbols $\Gamma \supset (O \cup \{\mathsf{A}, \mathsf{B}\})$.

Question: Does the existential team have a forced win in the following game? Players $\forall$ (universal), $\exists_1$, and $\exists_2$ (existential) take turns in that order, beginning with $\forall$. S's tape is initially set empty. On $\exists_i$'s turn, he makes a move from O. On $\forall$'s turn, he takes the following steps:

1. If not the first turn, record $\exists_1$'s and $\exists_2$'s moves in particular reserved cells of S's tape.

2. Simulate S using its current tape state as input, either until it terminates, or for k steps. If S accepts, $\forall$ wins the game. If S rejects, $\forall$ loses the game. Otherwise, leave the current contents of the tape as the next turn's input.

3. Make a move $(x, y) \in \{\mathsf{A}, \mathsf{B}\} \times \{\mathsf{A}, \mathsf{B}\}$, and record this move in particular reserved cells of S's tape.

The state of S's tape is always private to $\forall$. Also, $\exists_1$ sees only the value of x, and $\exists_2$ sees only the value of y. The existential players also do not see each other's moves. The existential team wins if either existential player wins.

Theorem 7.3. *Team Computation Game is undecidable.*

Proof: We reduce from acceptance of a Turing machine on an empty input, which is undecidable. Given a TM M, we construct TM S as above so that when it is run, it verifies that the moves from the existential players form valid computation histories, with each successive character following in the selected history, A or B. It needs no nondeterminism to do this; all the necessary nondeterminism by $\forall$ is in the moves (x, y). The $\exists$-options O are the tape alphabet of $M \cup \#$.

S maintains several state variables on its tape that are reused the next time it is run. First, it detects when both existential players are simultaneously beginning new configurations (by making move $\#$), for each of the four history pairs $\{A, B\} \times \{A, B\}$. Using this information, it maintains state that keeps track of when the configurations match. Configurations partially match for a history pair when either both are beginning new configurations, or both partially matched on the previous time step, and both histories just produced the same symbol. Configurations exactly match when they partially matched on the previous time step and both histories just began new configurations (with $\#$).

S also keeps track of whether one existential player has had one of its histories advanced exactly one configuration relative to one of the other player's histories.[3] It does this by remembering that two configurations exactly matched, and since then only one history of the pair has advanced, until finally it produced a $\#$. If one history in a history pair is advanced exactly one configuration, then this state continues as long as each history in the pair is advanced on the same turn. In this state, the histories may be checked against each other, to verify proper head motion, change of state, etc., by only remembering (on preserved tape cells) a finite number of characters from each history. S is designed to reject whenever this check fails, or whenever two histories exactly match and nonmatching characters are generated, and to accept when one computation history completes a configuration that is accepting for M. All of these computations may be performed in a constant number of steps; we use this number for k.

For any game history of A/B requests seen by $\exists_1$ ($\exists_2$), there is always some possible history of requests seen by $\exists_2$ ($\exists_1$) such that either $\exists_1$ ($\exists_2$) is on the first configuration (which must be empty), or $\exists_2$ ($\exists_1$) may have

---

[3]The number of steps into the history does not have to be exactly one configuration ahead; because M is deterministic, if the configurations exactly matched then one can be used to check the other's successor.

one of its histories exactly one configuration behind the currently requested history. Therefore, correct histories must always be generated to avoid losing.[4] Also, if correct accepting histories are generated, then the existential team will win, and thus the existential team can guarantee a win if and only if M accepts the empty string.   □

Next we define a team game played on Boolean formulas, and reduce Team Computation Game to this formula game. Traditionally one defines a formula game in a form for which it is easy to prove a hardness result, then reduces to another formula game with a cleaner definition and nicer properties. In this case, however, our formula game will only serve as an intermediate step on the way to a constraint-logic game, so no effort is made to define the simplest possible team formula game. On the contrary, the structure of the game is chosen so as to enable the simplest possible reduction to a constraint-logic game.

The reduction from Team Computation Game works by creating formulas that simulate the steps of Turing machine S.

TEAM FORMULA GAME

Instance: Sets of Boolean variables X, X', Y_1, Y_2; Boolean variables $h_1, h_2 \in X$; and Boolean formulas $F(X, X', Y_1, Y_2)$, $F'(X, X')$, and $G(X)$, where F implies $\overline{F'}$.

Question: Does White have a forced win in the following game? The six steps taken on each turn repeat in the following order:

1. B sets variables X to any values. If F and G are then true, Black wins.

2. If F is false, White wins. Otherwise, W_1 does nothing.

3. W_2 does nothing.

4. B sets variables X' to any values.

---

[4]Note that this fact depends on the nondeterminism of $\forall$ on each move. If instead $\forall$ followed a strategy of always advancing the same history pair, until it nondeterministically decided to check one against the other by switching histories on one side, the existential players could again gain information enabling them to cheat. This is a further difference from the original computation game from [133], where such a strategy is used; the key here is that $\forall$ is always able to detect when the histories happen to be nondeterministically aligned, and does not have to arrange for them to be aligned in advance by some strategy that the existential players could potentially take advantage of.

5. If F' is false, White wins. W_1 sets variables Y_1 to any values.

6. W_2 sets variables Y_2 to any values.

B sees the state of all the variables; W_i only sees the state of variables Y_i and h_i.

Theorem 7.4. *Team Formula Game is undecidable.*

Proof: Given an instance of Team Computation Game, we create the necessary variables and formulas as follows.

F will verify that B has effectively run TM S for k steps, by setting X to correspond to a valid nonrejecting computation history for it. (This can be done straightforwardly with $O(k^2)$ variables; see, for example, [28].) F also verifies that the values of Y_i are equal to particular variables in X, and that a set of "input" variables $I \subset X$ are equal to corresponding variables X'. X' thus represents the output of the previous run of S.

G is true when the copies of the Y_i in X represent an illegal white move (see below), or when X corresponds to an accepting computation history for S.

F' is true when the values X' equal those of a set of "output" variables $O \subset X$. These include variables representing the output of the run of S, and also h_1, h_2. We can assume without loss of generality here that S always changes its tape on a run. (We can easily create additional tape cells and states in S to ensure this if necessary, without affecting the simulation.) As a result, F implies $\overline{F'}$, as required; the values of X' cannot simultaneously equal those of the input and the output variables in X.

The $\forall$ player's move $(x, y) \in \{\mathsf{A}, \mathsf{B}\} \times \{\mathsf{A}, \mathsf{B}\}$ is represented by the assignments to history-selecting variables h_1 and h_2; false represents A and true B. The $\exists$-options O correspond to the Y_i; each element of O has one variable in Y_i, so that W_i must move by setting one of the Y_i to true and the rest to false.

Then, it is clear that the rules of Team Formula Game force the players effectively to play the given Team Computation Game. □

TPCL Reduction. Finally, we are ready to complete the undecidability reduction for TPCL. The overall reduction from Team Formula Game is shown in Figure 7.2. Before proving its correctness, we first examine the subcomponents represented by boxes in the figure.

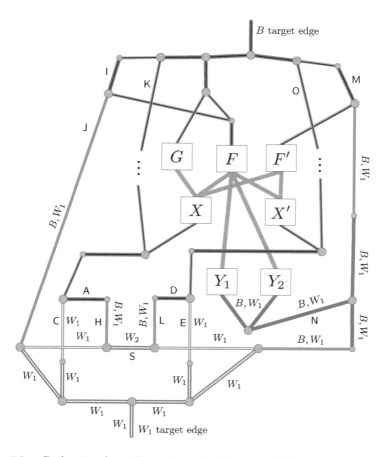

Figure 7.2.   Reduction from Team Formula Game to TPCL. White edges and multiplayer edges are labeled with their controlling player(s); all other edges are black. Thick gray lines represent bundles of black edges.

The F, F', and G boxes represent AND/OR subgraphs that implement the corresponding Boolean functions, as in earlier chapters. Their inputs come from outputs of the variable-set boxes. All these edges are black.

The boxes X and X' represent the corresponding variable sets. The incoming edge at the bottom of each box unlocks their values, by a series of latch gadgets (as in Section 5.2.1), shown in Figure 7.3(a). When the input edge is directed upward, the variable assignment may be freely changed; when it is directed down, the assignment is fixed.

The boxes Y_1 and Y_2 represent the white variables. An individual white variable for Player W_i is shown in Figure 7.3(b). B may activate the

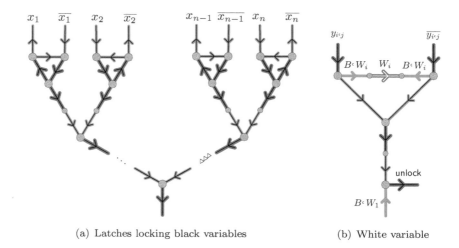

(a) Latches locking black variables   (b) White variable

Figure 7.3. Additional gadgets for TPCL reduction.

appropriate top output edge at any time; however, doing so also enables
the bottom output edge controlled jointly by B and W_1. If B wants to
prevent W_1 from directing this edge down, he must direct **unlock** right; but
then the black output edges are forced down, allowing W_i to freely change
the central variable edge. The **unlock** edges are left loose (shorthand for
using a free edge terminator); the bottom edges are ORed together to form
the single output edge for each box in Figure 7.2 (still jointly controlled
by B and W_1). Note that for variables controlled by W_2, W_1 can know
whether the variable is unlocked without knowing what its assignment is.

We will also consider some properties of the "switch" edge **S** before
delving into the proof. This edge is what forces the alternation of the two
types of B-W_1-W_2 move sequences in Team Formula Game. When **S** points
left, B is free to direct the connecting edges so as to unlock variables X.
But if B leaves edge **A** pointing left at the end of his turn, then W_1 can
immediately win, starting with edge **C**. (We skip label **B** to avoid confusion
with the B player label.) Similarly, if **S** points right, B can unlock the
variables in X', but if he leaves edge **D** pointing right, then W_1 can win
beginning with edge **E**. Later we will see that W_2 must reverse **S** each turn,
forcing distinct actions from B for each direction.

Theorem 7.5. *TPCL is undecidable, even with $N = 3$ players.*

Proof: Given an instance of Team Formula Game, we construct a TPCL
graph as described above. B sees the states of all edges; W_i sees only
the states of the edges he controls, those immediately adjacent (so that

he knows what moves are legal), and the edge in X corresponding to variable h_i.

We will consider each step of Team Formula Game in turn, showing that each step must be mirrored in the TPCL game. Suppose that initially, S points left. Here are the six steps:

1. B may set variables X to any values by unlocking their controlling latches, beginning with edge H. He may also direct the edges corresponding to the current values of X, Y_1, Y_2, and X' into formula networks F, F', and G, but he may not change the values of X', because their latches must be locked if S points left. If these moves enable him to satisfy formulas F and G, then he wins. Otherwise, if F is true, he may direct edge I upward. He must finish by redirecting A right, thus locking the X variables; otherwise, W_1 could then win as described above. Also, B may leave the states of Y_1 and Y_2 locked.

   B does not have time both to follow the above steps and to direct edge K upward within k moves; the pathway from H through "..." to K has $k - 3$ edges.

   Also, if F is true then M must point down at the end of B's turn, because F and F' cannot simultaneously be true.

2. If F is false, then I must point down. This will enable W_1 to win, beginning with edge J (because S still points left). Also, if H still points up, W_1 may direct it down, unlocking S; as above, A must point right. Otherwise W_1 has nothing useful to do. He may direct the bottom edges of the Y_1 variables downward, but nothing is accomplished by this, because S points left.

3. On this step W_2 has nothing useful to do but direct S right, which he must do. Otherwise...

4. If S still points left, then B can win, by activating the long series of edges leading to K; I already points up, so unlike in step 1, he has time for this.

   Otherwise, B can now set variables X' to any values, by unlocking their latches, beginning with edge L. If G was not true in step 1, then it cannot be true now, because X has not changed, so B cannot win that way. If F' is true, then he may direct edge M upward. Also, at this point B should unlock Y_1 and Y_2, by directing his output edges back in and activating the unlock edges in the white variable gadgets. This forces I down, because F depends on the Y_i.

   As in step 1, B cannot win by activating edge O, because he does not have time both to follow the above steps and to reach O within k

moves. (Note that M must point down at the beginning of this turn; see step 1.)

5. If any variable of Y_1 or Y_2 is still locked, W_1 can win by activating the pathway through N. Also, if F' is false then M must point down; this lets W_1 win. (In both cases, note that S points right.) Otherwise, W_1 may now set Y_1 to any values.

6. W_2 may now set Y_2 to any values. Also, W_2 must now direct S left again. If he does not, then on B's next turn he can win by activating O.

Thus, all players are both enabled and required to mimic the given Team Formula Game at each step, and so the White team can win the TPCL game if and only if it can win the Team Formula Game. □

7.2.2 Restricted Problem

As usual, we strengthen Theorem 7.5 to apply to planar graphs that use a restricted set of vertex types.

Theorem 7.6. *TPCL is undecidable, even for planar graphs that use only AND and OR vertices.*

Proof: All crossings in the reduction involve only edges controlled by a single player; we can replace these with crossover gadgets from Section 5.2.2 without changing the game. As usual, we can eliminate red-blue edges and loose edges with the techniques of Section 2.3. All remaining vertices are ANDs and ORs. (However, note that several different AND- and OR-subtypes are used in the reduction, which we do not enumerate.) □

Perspectives on Part I

In this chapter we step back and take a broader perspective on the results in the preceding chapters. We may view the family of constraint-logic games, taken as a whole, as a *hierarchy of complete problems*; this idea is developed in Section 8.1. We return to the overall theme of games as computation, this time from a more philosophical and speculative perspective, in Section 8.2. There we address the apparently nonsensical result from Chapter 7 that play in a game of fixed physical size can emulate a Turing machine with an infinite tape.

8.1   Hierarchies of Complete Problems

In 1976, Galil [70] proposed the notion of a *hierarchy of complete problems* and gave several examples. The concept is based on the observation that there are many problems known complete for the classes

$$L \subseteq NL \subseteq P \subseteq NP \subseteq PSPACE,$$

but at the time, there were no examples in the literature of quintuples of complete problems, one for each class, such that each problem was a restricted version of the next. Galil presents several such hierarchies, from domains of graph theory, automata theory, theorem proving, and games.[1]

---

[1] These games are all zero-player games, or simulations, in our sense of "game." They are loosely based on Conway's Game of Life.

Galil's definition of a hierarchy of complete problems is specific to those particular complexity classes. However, it seems reasonable to apply the same concept more broadly. In the current case, the family of constraint-logic games forms what could be considered to be a two-dimensional hierarchy of complete problems. The complexities of the games, ranging from zero-player to team games horizontally and bounded vs. unbounded vertically, stand in the following relation (see Figure I.1):

$$
\begin{array}{ccccccc}
\text{PSPACE} & \subseteq & \text{PSPACE} & \subseteq & \text{EXPTIME} & \subseteq & \text{RE} \\
\cup| & & \cup| & & \cup| & & \cup| \\
\text{P} & \subseteq & \text{NP} & \subseteq & \text{PSPACE} & \subseteq & \text{NEXPTIME}
\end{array}
$$

Furthermore, in each case a constraint-logic game is a restricted version of the game to its right or top. Starting with Team Private Constraint Logic, restricting the number of players to be two, the set of private edges to be empty, k (number of moves per turn) to 1, and requiring that the players control disjoint sets of edges yields Two-Player Constraint Logic. Restricting further so that one player has no edges to control[2] yields Nondeterministic Constraint Logic. Restricting further so that the sequence of moves is forced gives Deterministic Constraint Logic. (Technically this last step is not a proper restriction, but we can suppose that there is a general move-order constraint in the other games that is taken to allow any order by default.)

Similarly, adding the restriction that each edge may reverse at most once turns each of those games into their bounded counterparts. Finally, requiring that the graph be planar changes the complexity from P-complete to NC^3-easy in the case of Bounded Deterministic Constraint Logic.

8.2   Games, Physics, and Computation

In Section 7.2, on team games with private information, we showed that there are space-bounded games that are undecidable. At first, this fact seems counterintuitive and bizarre. There are only finitely many positions in the game, and yet somehow the state of the game must effectively encode an arbitrarily long Turing machine tape. How can this be? Eventually the state would have to repeat.

The short answer is that yes, the position must eventually repeat, but some of the players will not know when it is repeating. The entire history of the game is relevant to correct play, and of course the history grows with each move. So in a sense, the infinite tape has merely been shoved

---

[2]Technically each player must have a target edge, but it is easy to construct instances where one player effectively can do nothing, and the question is whether the other player can win. These are effectively Nondeterministic Constraint Logic games.

under the rug. However, the important point is that the infinite space has been *taken out of the definition of the problem*. Perhaps these games can only be played perfectly by players with infinite memories; perhaps not. That question depends on the nature of the players; perhaps they have access to some nonalgorithmic means of generating moves. In any case, the composition and capabilities of the players are not part of the definition of the problem—of the finite computing machine that is a game. A player is simply a black box which is fed game state and generates moves.

Compare the situation to the notion of a nondeterministic Turing machine. The conventional view is that a nondeterministic Turing machine is allowed to "guess" the right choice at each step of the computation. There is no question or issue of how the guess is made. Yet, one speaks of nondeterministic computations being "performed" by the machine. It is allowed access to a nonalgorithmic resource to perform its computation. Nondeterministic computers may or may not be "magical" relative to ordinary Turing machines; it is unknown whether $P = NP$. However, one kind of magic they definitely cannot perform is to turn finite space into infinite space. But a team-game "computer," on the other hand, *can* perform this kind of magic, using only a slight generalization of the notion of nondeterminism.

Whether these games can be played perfectly in the real world—and thus, whether we can actually perform arbitrary computations with finite physical resources—is a question of physics, not of computer science. And it is not immediately obvious that the answer must be no.

Others have explored various possibilities for squeezing unusual kinds of computation out of physical reality. There have been many proposals for how to solve NP-complete problems in polynomial time; Aaronson [1] offers a good survey. One such idea, which works if one subscribes to Everett's relative-state interpretation of quantum mechanics [54] (popularly called "many worlds"), is as follows. Say you want to solve an instance of SAT, which is NP-complete. You need to find a variable assignment that satisfies a Boolean formula with n variables. Then you can proceed as follows: guess a random variable assignment, and if it does not happen to satisfy the formula, kill yourself. Now, in the only realities you survive to experience you will have "solved" the problem in polynomial time.[3] Aaronson has termed this approach "anthropic computing."

---

[3]To avoid the problem with what happens when there is no satisfying assignment, Aaronson proposes you instead kill yourself with probability $1 - 2^{-2n}$ if you do not guess a satisfying assignment. Then if you survive without having guessed an assignment, it is almost certain that there is no satisfying assignment. This step is not strictly necessary, however. There would always be some reality in which you somehow avoided killing yourself; perhaps your suicide machine of choice failed to operate in some highly improbable way. Of course, for the technique to work at all, such a failure must be very improbable.

Apart from the possibly metaphysical question of whether there would indeed always be a "you" that survived this "computation," there is the annoying practical problem that those around you would almost certainly experience your death, instead of your successful efficient computation. There is a way around this problem, however. Suppose that, instead of killing yourself, you destroy the entire universe. Then, effectively, the entire universe is cooperating in your computation, and nobody will ever experience you failing and killing yourself. A related idea was explored in the science-fiction story "Doomsday Device" by John Gribbin [81]. In that story a powerful particle accelerator seemingly fails to operate, for no good reason. Then a physicist realizes that if it were to work, it would effectively destroy the entire universe, by initiating a transition from a cosmological false-vacuum state to a lower-energy vacuum state. In fact, the accelerator *has* worked; the only realities the characters experience involve highly unlikely equipment failures. (Whether such a false-vacuum collapse is actually possible is an interesting question [162].) We can imagine incorporating such a particle accelerator in a computing machine. We would like to propose the term "doomsday computation" for any kind of computation in which the existence of the universe might depend on the output of the computation. Clearly, doomsday computation is a special case of anthropic computation.

However, neither approach seems to offer the ability to perform *arbitrary* computations. Other approaches considered in [1] might do better: "time-travel computing," which works by sending bits along closed timelike curves (CTCs), can solve PSPACE-complete problems in polynomial time.

Perhaps there is some way to generalize some such "weird physics" kind of computation to enable perfect game play. The basic idea of anthropic computation seems appropriate: filter out the realities in which you lose, post-selecting worlds in which you win. But directly applied, as in the SAT example above, this only works for bounded one-player puzzles. Computing with CTCs gets you to PSPACE, which is suggestive of solving a two-player, bounded-length game, or a one-player, unbounded-length puzzle. Perhaps just one step more is all that is needed to create a perfect team-game player, and thus a physically finite, but computationally universal, computer.

Games in Particular

Part II applies the results of Part I to particular games and puzzles, to prove them hard. The simplicity of many of the reductions strengthens the view that constraint logic is a general game model of computation. Each reduction may be viewed as the construction of a kind of computer, using the physics provided by the game components at hand. Especially useful is the fact that the hardness results for constraint logic hold even when the graphs are planar. Traditionally some sort of crossover gadget has often been required for game and puzzle hardness proofs, and these are often among the most difficult gadgets to design.

For all of these results, it must be borne in mind that it is the *generalized* version of a game that is shown hard. For example, Amazons is typically played on a 10×10 board. But it is meaningless to discuss the complexity of a problem for a fixed input size; it is Amazons played on an $n \times n$ board that is shown PSPACE-complete. The "P" in "PSPACE" must be polynomial in something.

We give new hardness results for ten games: TipOver, Hitori, sliding-block puzzles, sliding-coin puzzles, plank puzzles, Push-2-F, hinged polygon dissections, Amazons, Konane, and Cross Purposes. Among these, sliding-block puzzles, Amazons, and Konane had been well-known open problems receiving study for some time.

We strengthen the existing hardness results for two games, the Warehouseman's Problem and Sokoban, and we give a simpler hardness proof than the extant one for Rush Hour, and show that a triangular version of Rush Hour is also hard.

One-Player Games (Puzzles)

In this chapter we present several new results for one-player games.

9.1 TipOver

TipOver is a puzzle in which the goal is to navigate a layout of vertical crates, tipping some over to reach others, so as to eventually reach a target crate. Crates can only tip into empty space, and you cannot jump over empty space to reach other crates. The challenge is to tip the crates in the right directions and the right order.

TipOver originated as an online puzzle created by James Stephens, called the "The Kung Fu Packing Crate Maze" [156]. Now it also exists in physical form (shown in Figure 9.1), produced by ThinkFun, the makers of Rush Hour and other puzzles. Like Rush Hour, TipOver comes with a board and a set of pieces, and 40 challenge cards, each with a different puzzle layout.

The standard TipOver puzzles are laid out on a 6×6 grid, but the puzzle naturally generalizes to $n \times n$ layouts. This is a bounded-move puzzle—each crate can only tip over once. Therefore, it is a candidate for a Bounded NCL reduction. We give a reduction showing that TipOver is NP-complete. (See also [90].)

Figure 9.1. TipOver puzzle. (Courtesy of ThinkFun, Inc.)

Rules. In its starting configuration, a TipOver puzzle has several vertical crates of various heights $(1 \times 1 \times h)$ arranged on a grid, with a "tipper"—representing a person navigating the layout—standing on a particular starting crate. There is a unique red crate, $1 \times 1 \times 1$, elsewhere on the grid; the goal is to move the tipper to this target red crate.

The tipper can tip over any vertical crate that it is standing on, in any of the four compass directions, provided that there is enough empty space within the grid for that crate to fall unobstructed and lie flat. The tipper is nimble enough to land safely on the newly fallen crate. The tipper can also walk, or climb, along the tops of any crates that are directly adjacent, even when they have different heights. However, the tipper is not allowed to jump empty space to reach another crate. It cannot even jump to a diagonally neighboring crate; the crates must be touching.

A sample puzzle and its solution are shown in Figure 9.2. The first layout is the initial configuration, with the tipper's location marked with a red square outline, and the height of each vertical crate indicated. In each successive step, one more crate has been tipped over.

9.1.1 NP-completeness

We reduce Bounded NCL (Section 5.1) to TipOver to show NP-hardness. Given an instance of Bounded NCL, we construct a TipOver puzzle that can be solved just when the target edge can be reversed. We need to build AND, OR, FANOUT, and CHOICE gadgets, as in Section 5.1.2, and show how to wire them together. We also need to build a single loose edge, as in Section 5.1.3, but this just corresponds to the tipper's starting point.

Figure 9.2. A sample TipOver puzzle and its solution.
(Courtesy of ThinkFun, Inc.)

All of our gadgets will be built with initially vertical, height-two crates. The mapping from constraint-graph properties to TipOver properties is that an edge can be reversed just when a corresponding TipOver region is reachable by the tipper.

One-Way Gadget. We will need an auxiliary gadget that can only be traversed by the tipper in one direction initially. The gadget, and the sequence of steps in a left-right traversal, are shown in Figure 9.3. Once it has been so traversed, a one-way gadget can be used as an ordinary wire. But if it is first approached from the right, there is no way to bridge the gap and reach the left side.

We will attach one-way gadgets to the inputs and outputs of each of the following gadgets.

OR/FANOUT Gadget. A simple intersection, protected with one-way gadgets, serves as an OR, shown in Figure 9.4(a).

Lemma 9.1. *The construction in Figure 9.4(a) satisfies the same constraints as a Bounded NCL OR vertex, with A and B corresponding to the input edges, and C corresponding to the output edge.*

Proof: The tipper can clearly reach C if and only if it can reach either A or B. Since the output is protected with a one-way gadget, the tipper cannot reach C by any other means. □

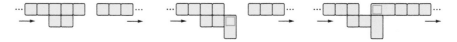

Figure 9.3. A wire that must be initially traversed from left to right. All crates are height two.

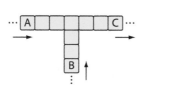

(a) OR gadget. If the tipper can reach either A or B, then it can reach C.

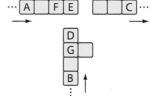

(b) AND gadget. If the tipper can reach both A and B, then it can reach C.

Figure 9.4. TipOver AND and OR gadgets.

Clearly, changing the direction of the one-way gadget protecting input A turns it input an output, and turns an OR gadget into a FANOUT gadget with the input at B.

AND Gadget. AND is a bit more complicated. The construction is shown in Figure 9.4(b). This time the tipper must be able to exit to the right only if it can independently enter from the left and from the bottom. This means that, at a minimum, it will have to enter from one side, tip some crates, retrace its path, and enter from the other side. Actually, the needed sequence will be a bit longer than that.

Lemma 9.2. *The construction in Figure 9.4(b) satisfies the same constraints as a Bounded NCL AND vertex, with A and B corresponding to the input edges, and C corresponding to the output edge.*

Proof: We need to show that the tipper can reach C if and only if it can first reach A and B. First, note that F is the only crate that can possibly be tipped so as to reach C; no other crate will do. If the tipper is only able to enter from A, and not from B, it can never reach C. The only thing that can be accomplished is to tip crate F down, so as to reach B from the wrong direction. But this does not accomplish anything, because once F has been tipped down it can never be tipped right, and C can never be reached. Suppose, instead, the tipper can enter from B, but not from A. Then again, it can reach A from the wrong direction, by tipping crate D

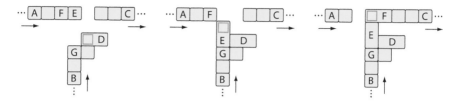

Figure 9.5. How to use the AND gadget.

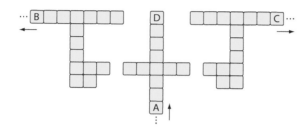

Figure 9.6. TipOver CHOICE gadget. If the tipper can reach A, then it can reach B or C, but not both.

right and crate G up. But again, nothing is accomplished by this, because now crate E cannot be gotten out of the way without stranding the tipper.

Now suppose the tipper can reach both A and B. Then the following sequence (shown in Figure 9.5) lets it reach C. First the tipper enters from B, and tips crate D right. Then it retraces its steps along the bottom input, and enters this time from A. Now it tips crate E down, connecting back to B. From here it can again exit via the bottom, return to A, and finally tip crate F right, reaching C. The right side winds up connected to the bottom input, so that the tipper can still return to its starting point as needed from later in the puzzle.   □

CHOICE Gadget. Finally, we need a CHOICE gadget. This is shown in Figure 9.6.

Lemma 9.3. *The construction in Figure 9.6 satisfies the same constraints as a Bounded NCL CHOICE vertex, with A corresponding to the input edge, and B and C corresponding to the output edges.*

Proof: Because of the built-in one-way gadgets, the only way the tipper can exit the gadget is by irreversibly tipping D either left or right. It may then reconnect to A by using the appropriate one-way pathway, but it can never reach the other side.   □

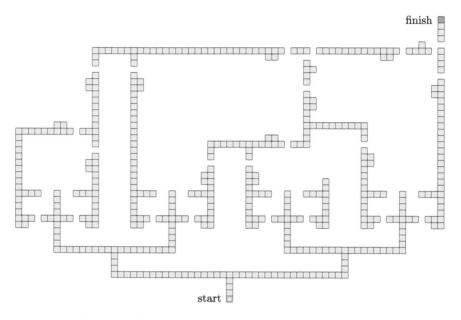

Figure 9.7. TipOver puzzle for a simple constraint graph.

Theorem 9.4. *TipOver is NP-complete.*

Proof: Reduction from Bounded NCL. Given a planar constraint graph made of AND, OR, FANOUT, CHOICE, and red-blue vertices, and with a single edge that may initially reverse, we construct a corresponding TipOver puzzle, as described above. The wiring connecting the gadgets together is simply a chain of vertical, height-2 crates. The tipper starts on some gadget input corresponding to the location of the single loose edge, and can reach the target crate just when the target edge in the constraint graph may be reversed. Therefore, TipOver is NP-hard.

TipOver is clearly in NP: there are only a linear number of crates that may tip over, and therefore a potential solution may be verified in polynomial time.   □

9.2   Hitori

The puzzle *Hitori* was popularized by Japanese publisher Nikoli, along with its more-famous sibling Sudoku, and several other "pencil-and-paper" puzzles—many of which have been shown NP-complete (see Appendix A).

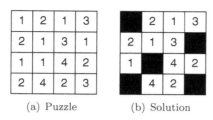

Figure 9.8. A simple Hitori puzzle and its solution.

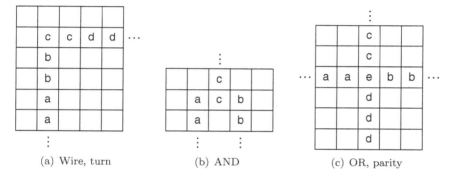

(a) Wire, turn   (b) AND   (c) OR, parity

Figure 9.9. Hitori gadgets.

In Hitori, we are given a rectangular grid with each square labeled with an integer, and the goal is to paint a subset of the squares so that (1) no row or column has a repeated unpainted label (similar to Sudoku), (2) painted squares are never (horizontally or vertically) adjacent, and (3) the unpainted squares are all connected (via horizontal and vertical connections). A simple Hitori puzzle and its solution are shown in Figure 9.8.

Hitori is a bounded one-player game, so we should expect that it could be NP-complete. We give a reduction from Constraint Graph Satisfiability (Section 5.1.3) showing NP-completeness.

9.2.1 NP-completeness

Given a planar constraint graph made of ANDs and ORs, we construct a Hitori puzzle that can be solved just when the constraint graph has a legal configuration. We need to build AND and OR gadgets, and show how to wire them together into a planar graph. The various gadgets we shall need are shown in Figure 9.9.

Wiring. We represent edge orientation with wires, or strings of adjacent squares, consisting of integers $x_1, x_1, x_2, x_2, ..., x_{n-1}, x_{n-1}, x_n, x_n$, where the x_i are distinct. If the first x_1 is unpainted, then the next must be painted (by rule 1 above), forcing the first x_2 to be unpainted (by rule 2), etc.; thus the last x_n must be painted. If the first x_1 is painted, the last x_n may be painted or unpainted (as far as this wire's constraints are concerned). The reason is that we could (for example) have the second x_1 and the first x_2 both unpainted without violating the rules.

Wires may be turned, as in Figure 9.9(a): if the bottom a is unpainted, then the right d must be painted. Here a, b, etc. are distinct integers. We assume that the unlabeled squares all contain distinct integers not otherwise used in the gadgets.

AND *Vertex.* In Figure 9.9(b), suppose that the lower a (or b) is unpainted. Then the other a (or b) must be painted (rule 1), forcing the lower c to be unpainted (rule 2), and the upper c to be painted. But if both the lower a and the lower b are painted, then the upper ones must be unpainted, allowing the lower c to be painted and the upper c to be unpainted. Similarly, if the upper c is unpainted, then the lower a and b must be painted. But if the upper c is painted, then the lower a and b may be painted or unpainted.

These are the same constraints that an AND vertex has: a painted "input" square (lower a or b) corresponds to an inward-directed red edge; an unpainted "output" square (upper c) corresponds to an outward-directed blue edge.

OR *Vertex/Parity Gadget.* In Figure 9.9(c), first, consider the ds. At most one can be unpainted, but no two adjacent may be painted. Therefore, both the lower and the upper one must be painted, and e must be unpainted.

Suppose that both the left a and the right b are unpainted. Then the right a and the left b must be painted. As an unpainted square, e must be connected to the other unpainted squares (rule 3); the lower c is the only way out. Therefore, the lower c is unpainted, and the upper one painted. But if either the left a or the right b is painted, then the other a or b will be unpainted, allowing another way out for e. Then, the lower c may be painted, and the upper c unpainted. These are the same constraints an OR vertex has, again with an unpainted "port" square (left a, right b, top c) corresponding to an outward-directed edge, and a painted port square corresponding to an inward-directed edge.

This gadget can also serve to alter the positional parity in wiring, so that the various gadgets can be connected arbitrarily, by using only one input, and blocking the other one (for example, by adding another b to the right of the right one).

Theorem 9.5. *Hitori is NP-complete.*

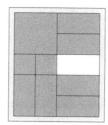

Figure 9.10. Dad's Puzzle.

Proof: Given a planar AND/OR constraint graph, we construct a Hitori puzzle by connecting together AND and OR vertex gadgets with wires, adjusting positional parity as needed. If the graph has a legal configuration, then every wire can be painted so as to satisfy all the Hitori constraints, as described. Similarly, if the Hitori puzzle can be solved, then a legal graph configuration can be read off the wires.

Hitori is clearly in NP: a potential solution may be verified in polynomial time.   □

9.3   Sliding-Block Puzzles

Sliding-block puzzles have long fascinated aficionados of recreational mathematics. From the infamous 15 Puzzle [152] associated with Sam Loyd to the latest whimsical variants such as Rush Hour, these puzzles seem to offer a maximum of complexity for a minimum of space.

In the usual kind of sliding-block puzzle, one is given a box containing a set of rectangular pieces, and the goal is to slide the blocks around so that a particular piece winds up in a particular place. A popular example is Dad's Puzzle, shown in Figure 9.10; it takes 59 moves to slide the large square to the bottom left.

Effectively, the complexity of determining whether a given sliding-block puzzle is solvable was an open problem for nearly 40 years. Martin Gardner devoted his February, 1964 Mathematical Games column to sliding-block puzzles. This is what he had to say [71, p. 65]:

> These puzzles are very much in want of a theory. Short of trial and error, no one knows how to determine if a given state is obtainable from another given state, and if it is obtainable, no one knows how to find the minimum chain of moves for achieving the desired state.

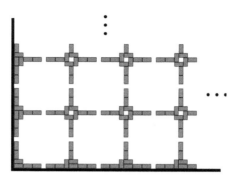

Figure 9.11. Sliding Blocks layout.

The computational complexity of sliding-block puzzles was considered explicitly by Spirakis and Yap in 1983 [154]; they showed that determining whether there is a solution to a given puzzle is NP-hard, and conjectured that it is PSPACE-complete. However, the form of the problem they considered was somewhat different from that framed here. In their version, the goal is to reach a given total configuration, rather than just moving a given piece to a given place, and there was no restriction on the sizes of blocks allowed. This problem was shown PSPACE-complete shortly afterwards, by Hopcroft, Schwartz, and Sharir [103], and renamed the *The Warehouseman's Problem*. (The Warehouseman's Problem is discussed in Section 9.4.)

This left the appropriate form of the decision question for most actual sliding-block puzzles open until we showed it PSPACE-complete in 2002 [84], based on the earlier result that the related puzzle Rush Hour is PSPACE-complete [56].

The *Sliding Blocks* problem is defined as follows: given a configuration of rectangles (*blocks*) of constant sizes in a rectangular two-dimensional box, can the blocks be translated and rotated, without intersection among the objects, so as to move a particular block?

9.3.1 PSPACE-completeness

We give a reduction from planar Nondeterministic Constraint Logic (NCL) showing that Sliding Blocks is PSPACE-hard even when all the blocks are 1×2 rectangles (dominoes). (Somewhat simpler constructions are possible if larger blocks are allowed; see Figure 1.2.) In contrast, there is a simple polynomial-time algorithm for 1×1 blocks; thus, the results are in some sense tight.

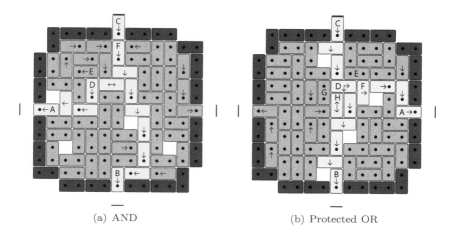

(a) AND   (b) Protected OR

Figure 9.12. Sliding Blocks vertex gadgets.

Sliding Blocks Layout. We fill the box with a regular grid of gate gadgets, within a "cell wall" construction as shown in Figure 9.11. The internal construction of the gates is such that none of the cell-wall blocks may move, thus providing overall integrity to the configuration.

AND and OR Vertices. We construct NCL AND and protected-OR (Section 5.2.3) vertex gadgets out of dominoes, in Figures 9.12(a) and 9.12(b). Each figure provides the bulk of an inductive proof of its own correctness, in the form of annotations. A dot indicates a square that is always occupied; the arrows indicate the possible positions a block can be in. For example, in Figure 9.12(b), block **D** may occupy its initial position, the position one unit to the right, or the position one unit down (but not, as we will see, the position one unit down and one unit right).

For each vertex gadget, if we inductively assume for each block that its surrounding annotations are correct, its own correctness will then follow, except for a few cases noted below. The annotations were generated by a computer search of all reachable configurations, but are easy to verify by inspection.

In each diagram, we assume that the cell-wall blocks (dark gray) may not move outward; we then need to show they may not move inward. The yellow ("trigger") blocks are the ones whose motion serves to satisfy the vertex constraints; the medium-gray blocks are fillers. Some of them may move, but none may move in such a way as to disrupt the vertices' correct operation.

The short lines outside the vertex ports indicate constraints due to adjoining vertices; none of the "port" blocks may move entirely out of its vertex. For it to do so, the adjoining vertex would have to permit a port block to move entirely inside the vertex, but in each diagram the annotations show this is not possible. Note that the port blocks are shared between adjoining vertices, as are the cell-wall blocks. For example, if we were to place a protected OR above an AND, its bottom port block would be the same as the AND's top port block.

A protruding port block corresponds to an inward-directed edge; a retracted block corresponds to an outward-directed edge. Signals propagate by moving "holes" forward. Sliding a block *out* of a vertex gadget thus corresponds to directing an edge *in* to a graph vertex.

Lemma 9.6. *The construction in Figure 9.12(a) satisfies the same constraints as an NCL AND vertex, with A and B corresponding to the AND red edges, and C to the blue edge.*

Proof: We need to show that block C may move down if and only if block A first moves left and block B first moves down.

First, observe that this motion is possible. The trigger blocks may each shift one unit in an appropriate direction, so as to free block C.

The annotations in this case serve as a complete proof of their own correctness, with one exception. Block D appears as though it might be able to slide upward, because block E may slide left, yet D has no upward arrow. However, for E to slide left, F must first slide down, but this requires that D first be slid down. So when E slides left, D is not in a position to fill the space it vacates.

Given the annotations' correctness, it is easy to see that it is not possible for C to move down unless A moves left and B moves down.   □

Lemma 9.7. *The construction in Figure 9.12(b) satisfies the same constraints as an NCL protected-OR vertex, with A and B corresponding to the protected edges.*

Proof: We need to show that block C may move down if and only if block A first moves right, or block B first moves down.

First, observe that these motions are possible. If A moves right, D may move right, releasing the blocks above it. If B moves down, the entire central column may also move down.

The annotations again provide the bulk of the proof of their own correctness. In this case there are three exceptions. Block E looks as if it might be able to move down, because D may move down and F may move right. However, D may only move down if B moves down, and F may only move right if A moves right. Because this is a protected OR, we are guaranteed

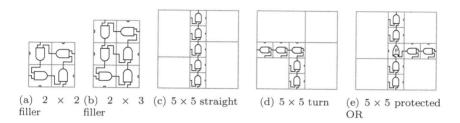

(a) 2 × 2 (b) 2 × 3 (c) 5 × 5 straight   (d) 5 × 5 turn   (e) 5 × 5 protected
filler   filler   OR

Figure 9.13. Sliding Blocks wiring.

that this cannot happen: the vertex will be used only in graphs such that at most one of A and B can slide out at a time. Likewise, G could move right if D were moved right while H were moved down, but again those possibilities are mutually exclusive. Finally, D could move both down and right one unit, but again this would require A and B to both slide out.

Given the annotations' correctness, it is easy to see that it is not possible for C to move down unless A moves right or B moves down.   □

Graphs. Now that we have AND and protected-OR gates made out of sliding-blocks configurations, we must next connect them together into arbitrary planar graphs. First, note that the box wall constrains the facing port blocks of the vertices adjacent to it to be retracted (see Figure 9.11). This does not present a problem, however, as we will show. The unused ports of both the AND and protected-OR vertices are unconstrained; they may be slid in or out with no effect on the vertices. Figures 9.13(a) and 9.13(b) show how to make (2 × 2)-vertex and (2 × 3)-vertex "filler" blocks out of ANDs. (We use conventional "and" and "or" icons to denote the vertex gadgets.) Because none of the ANDs need ever activate, all the exterior ports of these blocks are unconstrained. (The unused ports are drawn as semicircles.)

We may use these filler blocks to build (5 × 5)-vertex blocks corresponding to "straight" and "turn" wiring elements (Figures 9.13(c) and 9.13(d)). Because the filler blocks may supply the missing inputs to the ANDs, the "output" of one of these blocks may activate (slide in) if and only if the "input" is active (slid out). Also, we may "wrap" the AND and protected-OR vertices in 5 × 5 "shells," as shown for protected OR in Figure 9.13(e). (Note that "left turn" is the same as "right turn"; switching the roles of input and output results in the same constraints.)

We use these 5 × 5 blocks to fill the layout; we may line the edges of the layout with unconstrained ports. The straight and turn blocks provide the necessary flexibility to construct any planar graph, by letting us extend the vertex edges around the layout as needed.

Theorem 9.8. *Sliding Blocks is PSPACE-complete, even for* 1×2 *blocks.*

Proof: Reduction from NCL. Given a planar constraint graph made of AND and protected-OR vertices, we construct a corresponding sliding-block puzzle, as described above. A port block of a particular vertex gadget may move if and only if the corresponding NCL graph edge may be reversed.

Sliding Blocks is in PSPACE: a simple nondeterministic algorithm traverses the state space, as in Theorem 5.9.   □

9.4   The Warehouseman's Problem

As mentioned in Section 9.3, the *Warehouseman's Problem* is a particular formulation of a kind of sliding-block problem in which the blocks are not required to have a fixed size, and the goal is to put each block at a specified final position. Hopcroft, Schwartz, and Sharir [103] showed the Warehouseman's Problem PSPACE-hard in 1984.

Their construction critically requires that some blocks have dimensions that are proportional to the box dimensions. Using Nondeterministic Constraint Logic, we can strengthen (and greatly simplify) the result: it is PSPACE-complete to achieve a specified total configuration, even when the blocks are all 1×2.

9.4.1   PSPACE-completeness

Theorem 9.9. *The Warehouseman's Problem is PSPACE-hard, even for* 1×2 *blocks.*

Proof: As in Section 9.3, but using Theorem 5.15, which shows that determining whether a given total configuration may be reached from a given AND/OR graph is PSPACE-hard. The graph initial and desired configurations correspond to two block configurations; the second is reachable from the first if and only if the NCL problem has a solution.   □

If we restrict the block motions to unit translations (as appropriate when viewing the problem as a generalized combinatorial game), then the problem is also in PSPACE, as in Theorem 5.9.

9.5   Sliding-Coin Puzzles

Sliding-block puzzles have an obvious complexity about them, so it is no surprise that they are PSPACE-complete. What is more surprising is that there are PSPACE-complete sliding-coin puzzles. For sliding-block puzzles, if the blocks are all 1×1, as in the 15 Puzzle, the puzzles become easy—it is the fact that one block can be dependent on the positions of two other blocks for the ability to move in a particular direction that makes it possible to build complex puzzles and gadgets. In a typical sliding-coin puzzle, a coin is like a 1×1 block; it only needs one other coin to move out of the way for it to be able to move and take its place. Indeed, many forms of sliding-coin puzzle have been shown to be efficiently solvable [39].

But it turns out that adding a simple constraint to the motion of the coins leads to a very natural problem that is PSPACE-complete.

The *Sliding Tokens* problem is defined as follows. It is played on an undirected graph with tokens placed on some of the vertices. A *legal configuration* of the graph is a token placement such that no adjacent vertices both have tokens. (That is, the tokens form an independent set of vertices.) A move is made by sliding a token from one vertex to an adjacent one, along an edge, such that the resulting configuration is legal. Given an initial configuration, is is possible to move a given token?

Note that this problem is essentially a dynamic, puzzle version of the Independent Set problem, which is NP-complete [74]. Similarly, the natural two-player-game version of Independent Set, called Kayles, is also PSPACE-complete [74]. Just as many NP-complete problems become PSPACE-complete when turned into two-player games [147], it is also natural to expect that they become PSPACE-complete when turned into dynamic puzzles.

Finally, from a more computational perspective, sliding-token graphs also superficially resemble Petri nets.

9.5.1 PSPACE-completeness

We give a reduction from Nondeterministic Constraint Logic showing that this problem is PSPACE-complete. By Theorem 5.12, we must show how to construct planar graphs made from AND and OR vertices.

AND and OR Vertices. We construct NCL AND and OR vertex gadgets out of sliding-token subgraphs, in Figures 9.14(a) and 9.14(b). The edges that cross the dotted-line gadget borders are "port" edges. A token on an outer port-edge vertex represents an inward-directed NCL edge, and vice-versa. Given an AND/OR graph and configuration, we construct a corresponding sliding-token graph, by joining together AND and OR vertex gadgets at their shared port edges, placing the port tokens appropriately.

Theorem 9.10. *Sliding Tokens is PSPACE-complete.*

Proof: First, observe that no port token may ever leave its port edge. Choosing a particular port edge **A**, if we inductively assume that this condi-

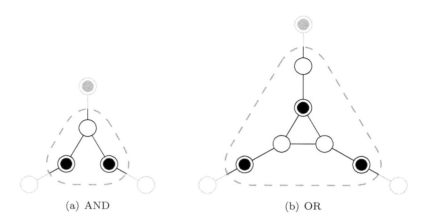

(a) AND   (b) OR

Figure 9.14. Sliding Tokens vertex gadgets.

tion holds for all other port edges, then there is never a legal move outside
A for its token; another port token would have to leave its own edge first.

The AND gadget clearly satisfies the same constraints as an NCL AND
vertex; the upper token can slide in just when both lower tokens are slid
out. Likewise, the upper token in the OR gadget can slide in when either
lower token is slid out; the internal token can then slide to one side or the
other to make room. It thus satisfies the same constraints as an NCL AND
vertex.

Sliding Tokens is in PSPACE: a simple nondeterministic algorithm traverses the state space, as in Theorem 5.9.   □

9.6   Plank Puzzles

A *plank puzzle* is a puzzle in which the goal is to cross a crocodile-infested
swamp, using only wooden planks supported by tree stumps.

Plank puzzles were invented by UK maze enthusiast Andrea Gilbert.
Like TipOver, they originated as a popular online puzzle applet [75]; now
there is also a physical version, sold by ThinkFun as River Crossing. Like
Rush Hour and TipOver, the puzzle comes with a set of challenge cards,
each with a different layout. Also like Rush Hour, and unlike TipOver,
plank puzzles are unbounded games; there is no resource that is used up as
the game is played. (By contrast, in TipOver, the number of vertical crates
must decrease by one each turn.) We give a reduction from Nondeterministic Constraint Logic showing that plank puzzles are PSPACE-complete.
(See also [87].)

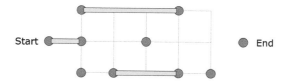

Figure 9.15. A plank puzzle.

Rules. The game board is an $n \times n$ grid, with *stumps* at some intersections, and *planks* arranged between some pairs of stumps, along the grid lines. The goal is to go from one given stump to another. You can pick up planks, and put them down between other stumps separated by exactly the plank length. You are not allowed to cross planks over each other, or over intervening stumps, and you can carry only one plank at a time.

A sample plank puzzle is shown in Figure 9.15. The solution begins as follows: walk across the length-1 plank; pick it up; lay it down to the south; walk across it; pick it up again; lay it down to the east; walk across it again; pick it up again; walk across the length-2 plank; lay the length-1 plank down to the east; ....

9.6.1   PSPACE-completeness

We give a reduction from Nondeterministic Constraint Logic. We need AND and OR vertices, and a way to wire them together to create a plank puzzle corresponding to any given planar graph.

The constraint-graph edge orientations are represented by the positions of "port planks" at each vertex interface; moving a port plank into a vertex gadget enables it to operate appropriately, and prevents it from being used in the paired vertex gadget.

AND vertex. The plank-puzzle AND vertex is shown in Figure 9.16(a). The length-2 planks serve as the input and output ports. (Of course, the gate may be operated in any direction.) Both of its input port planks (**A** and **B**) are present, and thus activated; this enables you to move its output port plank (**C**) outside the gate. Suppose you are standing at the left end of plank **A**. First walk across this plank, pick it up, and lay it down in front of you, to reach plank **D**. With **D** you can reach plank **B**. With **B** and **D**, you can reach **C**, and escape the gate. At the end of this operation **A** and **B** are trapped inside the vertex, inaccessible to the adjoining vertex gadgets.

The operation is fully reversible, since the legal moves in plank puzzles are reversible.

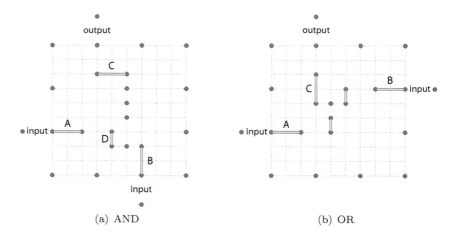

(a) AND   (b) OR

Figure 9.16.   Plank-puzzle AND and OR vertices.

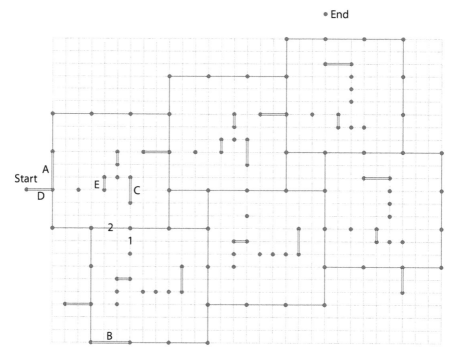

Figure 9.17.   A plank puzzle made from an AND/OR graph.

Figure 9.18. The equivalent constraint graph for Figure 9.17.

OR vertex. The OR vertex is shown in Figure 9.16(b). In this case, starting at either **A** or **B** will let you move the output plank **C** outside the vertex, by way of internal length-1 plank(s), trapping the starting plank inside the vertex.

Constraint Graphs. To complete the construction, we must have a way to wire these gates together into large puzzle circuits. Once you have activated an AND gate, you're stuck standing on its output plank—now what?

Figure 9.17 shows a puzzle made from six gates. For reference, the equivalent constraint graph is shown in Figure 9.18. The gates are arranged on a staggered grid, in order to make matching inputs and outputs line up. The port planks are shared between adjoining gates. Notice that two length-3 planks have been added to the puzzle. These are the key to moving around between the gates. If you are standing on one of these planks, you can walk along the edges of the gates, by repeatedly laying the plank in front of you, walking across it, then picking it up. This will let you get to any port of any of the gates. By using both the length-3 planks, you can alternately place one in front of the other, until you reach the next port you want to exit from. Then you can leave one length-3 plank there, and use the remaining one to reach the desired port entrance.

However, you cannot get inside any of the gates using just a length-3 plank, because there are no interior stumps exactly three grid units from a border stump.

To create arbitrary planar constraint graphs, we can use the same techniques used in Section 9.3 to build large "straight" and "turn" blocks out of 5 × 5 blocks of vertex gadgets.

Theorem 9.11. *Plank puzzles are PSPACE-complete.*

Proof: Reduction from NCL, by the construction described. A given stump may be reached if and only if the corresponding NCL graph edge may be reversed.

Plank puzzles are in PSPACE: a simple nondeterministic algorithm traverses the state space, as in Theorem 5.9.   □

9.7 Sokoban

In the pushing-blocks puzzle *Sokoban*, one is given a configuration of 1×1 blocks, and a set of target positions. One of the blocks is distinguished as the *pusher*. A move consists of moving the pusher a single unit either vertically or horizontally; if a block occupies the pusher's destination, then that block is pushed into the adjoining space, providing it is empty. Otherwise, the move is prohibited. Some blocks are *barriers*, which may not be pushed. The goal is to make a sequence of moves such that there is a (non-pusher) block in each target position.

By showing how to construct a Sokoban position corresponding to a space-bounded Turing machine, Culberson [33] proved that Sokoban is PSPACE-complete. Using Nondeterministic Constraint Logic, we give an alternate proof. Our result applies even if there are no barriers allowed in the Sokoban position, thus strengthening Culberson's result.

9.7.1 PSPACE-completeness

Unrecoverable Configurations. The idea of an *unrecoverable configuration* is central to Culberson's proof, and it will be central to our proof as well. We construct our Sokoban instance so that if the puzzle is solvable, then the original configuration may be restored from any solved state by reversing all the pushes. Then any push that may not be reversed leads to an unrecoverable configuration. For example, in the partial configuration in Figure 9.19(a), if block A is pushed left, it will be irretrievably stuck next to block D; there is no way to position the pusher so as to move it again. We may speak of such a move as being prohibited, or impossible, in the sense that no solution to the puzzle can include such a move, even though it is technically legal.

AND and OR Vertices. We construct NCL AND and OR vertex gadgets out of partial Sokoban positions, in Figure 9.19. (The pusher is not shown.) The gray blocks in the figures, though unmovable, are not barriers; they are simply blocks that cannot be moved by the pusher because of their configuration. The yellow "trigger" blocks are the ones whose motion serves to satisfy the vertex constraints. In each vertex, blocks A and B represent outward-directed edges; block C represents an inward-directed edge. A and C switch state by moving left one unit; B switches state by moving up one unit. We assume that the pusher may freely move to any empty space

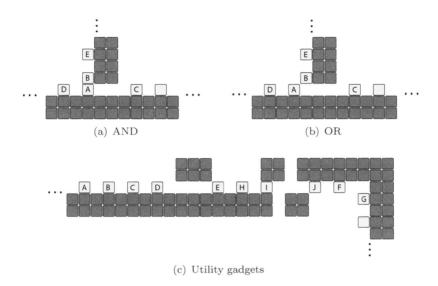

(a) AND

(b) OR

(c) Utility gadgets

Figure 9.19. Sokoban gadgets.

surrounding a vertex. We also assume that block D in Figure 9.19(a) may not reversibly move left more than one unit. Later, we show how to arrange both of these conditions.

Lemma 9.12. *The construction in Figure 9.19(a) satisfies the same constraints as an NCL AND vertex, with A and B corresponding to the AND red edges, and C to the blue edge.*

Proof: We need to show that C may move left if and only if A first moves left, and B first moves up. For this to happen, D must first move left, and E must first move up; otherwise pushing A or B would lead to an unrecoverable configuration. Having first pushed D and E out of the way, we may then push A left, B up, and C left. However, if we push C left without first pushing A left and B up, then we will be left in an unrecoverable configuration; there will be no way to get the pusher into the empty space left of C to push it right again. (Here we use the fact that D can only move left one unit.) □

Lemma 9.13. *The construction in Figure 9.19(b) satisfies the same constraints as an NCL OR vertex.*

Proof: We need to show that C may move left if and only if A first moves left, or B first moves up.

As before, D or E must first move out of the way to allow A or B to move. Then, if A moves left, C may be pushed left; the gap opened up by moving A lets the pusher get back in to restore C later. Similarly for B.

However, if we push C left without first pushing A left or B up, then, as in Lemma 9.12, we will be left in an unrecoverable configuration.   □

Graphs. We have shown how to make AND and OR vertices, but we must still show how to connect them up into arbitrary planar graphs. The remaining gadgets we shall need are illustrated in Figure 9.19(c).

The basic idea is to connect the vertices together with alternating sequences of blocks placed against a double-thick wall, as in the left of Figure 9.19(c). Observe that for block A to move right, first D must move right, then C, then B, then finally A, otherwise two blocks will wind up stuck together. Then, to move block D left again, the reverse sequence must occur. Such movement sequences serve to propagate activation from one vertex to the next.

We may switch the "parity" of such strings, by interposing an appropriate group of six blocks: E must move right for D to, then D must move back left for E to. We may turn corners: for F to move right, G must first move down. Finally, we may "flip" a string over, to match a required orientation at the next vertex, or to allow a turn in a desired direction: for H to move right, I must move right at least two spaces; this requires that J first move right.

We satisfy the requirement that block D in Figure 9.19(a) may not reversibly move left more than one unit by protecting the corresponding edge of every AND with a turn; observe that in Figure 9.19(c), block F may not reversibly move right more than one unit. The flip gadget solves our one remaining problem: how to position the pusher freely wherever it is needed. Observe that it is always possible for the pusher to cross a string through a flip gadget. (After moving J right, we may actually move I *three* spaces right.) If we simply place at least one flip along each wire, then the pusher can get to any side of any vertex.

Theorem 9.14. *Sokoban is PSPACE-complete, even if no barriers are allowed.*

Proof: Reduction from Nondeterministic Constraint Logic. Given a planar AND/OR graph, we build a Sokoban puzzle as described above, corresponding to the initial graph configuration. We place a target at every position that would be occupied by a block in the Sokoban configuration corresponding to the target graph configuration. Since NCL is inherently reversible, and our construction emulates NCL, then the solution configuration must also be reversible, as required for the unrecoverable configuration constraints.

Sokoban is in PSPACE: a simple nondeterministic algorithm traverses the state space, as in Theorem 5.9.   □

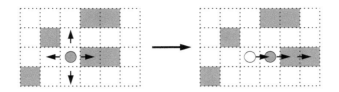

Figure 9.20. Example of pushing blocks.

9.8 Push-2-F

Push-2-F is a kind of block-pushing puzzle similar to the classic Sokoban (Section 9.7). The puzzle consists of unit square blocks on an integer lattice; some of the blocks are movable. (The "F" denotes that some blocks are fixed, and immovable.) A mobile "pusher" block may move horizontally and vertically in order to reach a specified goal position, and may push up to two blocks at once, as shown in Figure 9.20. More generally, in Push-k-F, the pusher may push up to k blocks. Unlike Sokoban, the goal is merely to get the pusher to a target location. This simpler goal makes constructing gadgets more difficult; in Sokoban, requiring multiple blocks to have target locations effectively prevents moves that could otherwise break the gadgets.

Here we present all of the gadgets needed to show that Push-k-F is PSPACE-complete for $k \geq 2$ (joint work with Michael Hoffmann), and state their properties. Some of the proofs that the gadgets satisfy the stated properties are rather detailed; we omit them here and refer the reader to [40] for details.

9.8.1 PSPACE-completeness

We reduce Nondeterministic Constraint Logic to Push-2-F to show PSPACE-completeness.

Basic Gadgets. In Figure 9.21 we show the basic Push-2-F gadgets. Red squares represent fixed blocks, and blue squares movable blocks.

The *diode* can be traversed arbitrarily often from I to O, but never from O to I.

The *join* can be traversed from both I_1 and I_2 to O arbitrarily often, but never from I_1 to I_2 or vice versa. Also, the pusher can always go from O to the most recently entered input (I_1 or I_2).

The *one-time-passage* gadget (shown semi-symbolically, with embedded diodes) allows the pusher to pass once from I to O and back to I (passage); after such a traversal, it must be reset during a $U \rightarrow U$ (unlock) traversal before another passage.

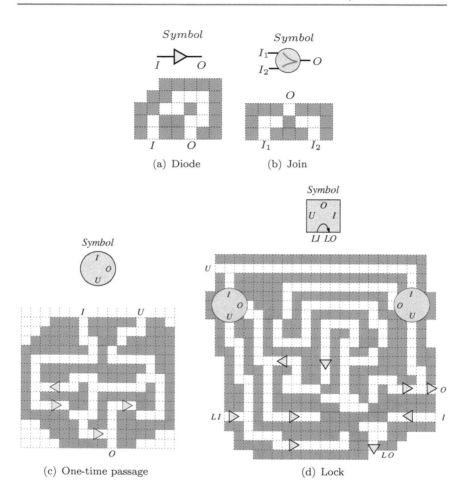

Figure 9.21. Basic Push-2-F gadgets. The one-time passage and lock gadgets are shown semi-symbolically; triangles denote embedded diodes, and labeled circles denote embedded one-time passage gadgets.

The *lock* gadget (shown semi-symbolically, with embedded diodes and one-time passage gadgets) can only be traversed via the following three "atomic" traversals: $LI \rightarrow LO$ (lock), $I \rightarrow O$ (passage), and $U \rightarrow U$ (unlock). Furthermore, any $LI \rightarrow LO$ and $I \rightarrow O$ traversals must be separated by a $U \rightarrow U$ traversal. This lock gadget is extremely complicated; a much simpler lock (which also does not require the one-time-passage gadget) is possible if the pusher is allowed to push three blocks instead of just two (Push-3-F). See [40] for details.

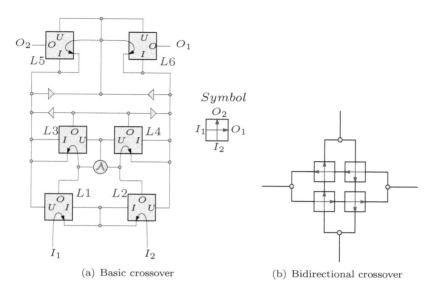

(a) Basic crossover   (b) Bidirectional crossover

Figure 9.22. Crossing gadgets.

Crossing Gadgets. From here on we represent gadgets schematically, in terms of previously defined gadgets, with "wires" representing corridors bounded by fixed blocks.

The *crossover* gadget shown in Figure 9.22(a) is made of six locks. It can be traversed from I_1 to O_1 and from I_2 to O_2 arbitrarily often; no other traversals are possible. The reader may wonder why a crossover gadget is necessary: don't we get crossovers for free with Nondeterministic Constraint Logic? The problem here is that we must cross signals even to build the constraint-logic gadgets themselves.

Using four crossovers, we can make a *bidirectional crossover*, shown in Figure 9.22(b).

AND and OR Vertices. We build the NCL vertex gadgets out of *buffered locks*. A buffered lock (Figure 9.23(a)) has the same properties as a lock, except that it may be unlocked during an $I \rightarrow O$ traversal, it may be unlocked by a $U \rightarrow E$ traversal, and the arrangement of terminals is different.

Each vertex gadget (Figures 9.23(b) and 9.23(c)) is made of three buffered locks, plus associated circuitry to enforce the necessary constraints. Each buffered lock acts as half of an edge: locked corresponds to "directed outward," and unlocked corresponds to "directed inward." The unattached buffered-lock terminals (E, some I's) are open to free space, which can reach any other such terminal, via appropriately placed crossing gadgets.

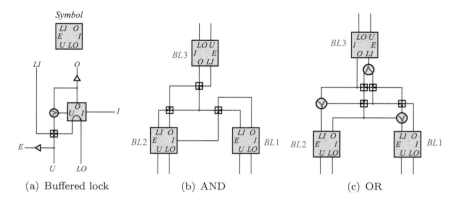

Figure 9.23. NCL vertex gadgets.

In each vertex, we assume that the lock states initially satisfy the vertex constraints. Then any possible pusher traversal maintains those constraints.

Lemma 9.15. *The gadget shown in Figure 9.23(b) satisfies the same constraints as an NCL AND vertex.*

Proof: To lock $BL3$, both $BL1$ and $BL2$ must be unlocked: the pusher may then traverse $I(BL1) \rightarrow LO(BL3)$, passing through $BL2$. To lock either $BL1$ or $BL2$, $BL3$ must be unlocked: the pusher may then traverse either $I(BL3) \rightarrow LO(BL1)$ or $I(BL3) \rightarrow LO(BL2)$. □

Lemma 9.16. *The gadget shown in Figure 9.23(c) satisfies the same constraints as an NCL OR vertex.*

Proof: Any buffered lock may be locked if and only if any other lock is unlocked. For example, if $BL1$ is unlocked, the pusher may traverse $I(BL1) \rightarrow LO(BL3)$; the other cases are symmetric. The join gadgets ensure that only the appropriate paths may be taken. □

Constraint Graphs. Vertices are connected together into arbitrary NCL constraint graphs by connecting the buffered locks in pairs, matching LO terminals to U terminals. Then any buffered lock can be unlocked precisely when its connecting buffered lock is locked. This property represents that a half-edge can be directed inwards precisely when the other half is directed outwards.

For example, suppose that $BL1$ and $BL2$ in an AND vertex are unlocked. Then the pusher may traverse $I(BL1) \rightarrow LO(BL3)$, and continue on to traverse $U \rightarrow E$ in the adjoining buffered lock.

Theorem 9.17. *Push-k-F is PSPACE-complete for $k \geq 2$.*

Proof: Reduction from NCL, by the construction described. A given NCL constraint graph may be represented as a Push-k-F configuration. The target edge in the NCL graph may be eventually reversed if and only if the pusher may reach the unlock terminal of a corresponding buffered lock.

Push-k-F is in PSPACE: a simple nondeterministic algorithm traverses the state space, as in Theorem 5.9. □

9.9   Rush Hour

In the puzzle *Rush Hour*, one is given a sliding-block configuration with the additional restriction that each block is constrained to move only horizontally or vertically on a grid. The goal is to move a particular block to a particular location at the edge of the grid. In the commercial version of the puzzle, the grid is 6×6, the blocks are all 1×2 or 1×3 ("cars" and "trucks"), and each block constraint direction is the same as its lengthwise orientation.

By showing how to build a kind of reversible computer from Rush Hour gadgets that work like constraint-logic AND and OR vertices, as well as a crossover gadget, Flake and Baum [56] showed that the generalized problem is PSPACE-complete. (Their construction was the basis for the development of constraint logic.) Tromp [164, 165] strengthened their result by showing that Rush Hour is PSPACE-complete even if the blocks are all 1×2.

Here we give a simpler construction showing that Rush Hour is PSPACE-complete, again using the traditional 1×2 and 1×3 blocks that must slide lengthwise. We only need an AND and a protected OR (Section 5.2.3), which turns out to be easier to build than OR; because of the generic crossover construction (Section 5.2.2), we don't need a crossover gadget. (We also don't need the miscellaneous wiring gadgets used in [56].)

9.9.1   PSPACE-completeness

Rush Hour Layout. We tile the grid with our vertex gadgets, as shown in Figure 9.24(a). One block (T) is the target, which must be moved to the bottom left corner; it is released when a particular port block slides into a vertex.

Dark-gray blocks represent the "cell walls," which unlike in our sliding-blocks construction are not shared. They are arranged so that they may not move at all. Yellow blocks are "trigger" blocks, whose motion serves to satisfy the vertex constraints. Medium-gray blocks are fillers; some of them may move, but they do not disrupt the vertices' operation.

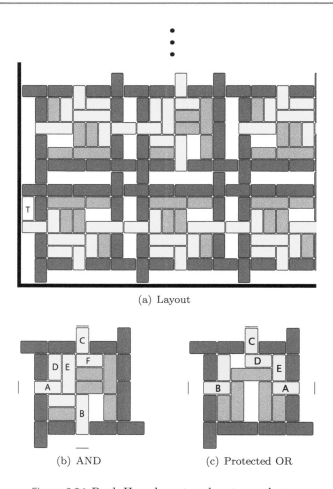

(a) Layout

(b) AND   (c) Protected OR

Figure 9.24. Rush Hour layout and vertex gadgets.

As in the sliding-blocks construction (Section 9.3), edges are directed inward by sliding blocks out of the vertex gadgets; edges are directed outward by sliding blocks in. The layout ensures that no port block may ever slide out into an adjacent vertex; this helps keep the cell walls fixed.

Lemma 9.18. *The construction in Figure 9.24(b) satisfies the same constraints as an NCL AND vertex, with A and B corresponding to the AND red edges, and C to the blue edge.*

Proof: We need to show that **C** may move down if and only if **A** first moves left and **B** first moves down.

Moving A left and B down allows D and E to slide down, freeing F, which releases C. The filler blocks on the right ensure that F may only move left; thus, the inputs are required to move to release the output.   □

Lemma 9.19. *The construction in Figure 9.24(c) satisfies the same constraints as an NCL protected-OR vertex, with A and B corresponding to the protected edges.*

Proof: We need to show that C may move down if either A first moves left or B first moves right.

If either A or B slides out, this allows D to slide out of the way of C, as required. Note that we are using the protected-OR property: if A were to move right, E down, D right, C down, and B left, we could not then slide A left, even though the OR property should allow this; E would keep A blocked. But in a protected OR we are guaranteed that A and B will not simultaneously be slid out.   □

Graphs. We may use the same constructions here that we used for sliding-blocks layouts: 5×5 blocks of Rush Hour vertex gadgets serve to build all the wiring necessary to construct arbitrary planar graphs (Figure 9.13).

In the special case of arranging for the target block to reach its destination, this will not quite suffice; however, we may direct the relevant signal to the bottom left of the grid, and then remove the bottom two rows of vertices from the bottommost 5×5 blocks; these can have no effect on the graph. The resulting configuration, shown in Figure 9.24(a), allows the target block to be released properly.

Theorem 9.20. *Rush Hour is PSPACE-complete.*

Proof: Reduction from NCL. Given a planar constraint graph made of AND and protected-OR vertices, we construct a corresponding Rush Hour puzzle, as described above. The output port block of a particular vertex may move if and only if the corresponding NCL graph edge may be reversed. We direct this signal to the lower left of the grid, where it may release the target block.

Rush Hour is in PSPACE: a simple nondeterministic algorithm traverses the state space, as in Theorem 5.9.   □

9.9.2 Generalized Problem Bounds

We may consider the more general *Constrained Sliding Blocks* problem, where blocks need not be 1×2 or 1×3, and may have a constraint direction independent of their dimension. In this context, the existing Rush Hour results do not yet provide a tight bound; the complexity of the problem for 1×1 blocks has not been addressed.

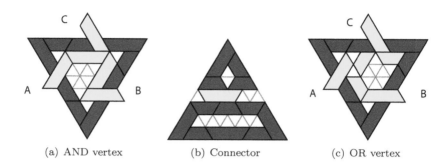

(a) AND vertex   (b) Connector   (c) OR vertex

Figure 9.25. Triagonal Slide-Out gadgets.

Deciding whether a block may move at all is in P: e.g., we may do a breadth-first search for a movable block that would ultimately enable the target block to move, beginning with the blocks obstructing the target block. Since no block need ever move more than once to free a dependent block, it is safe to terminate the search at already-visited blocks.

Therefore, a straightforward application of constraint logic cannot show this problem hard; however, the complexity of moving a given block to a given position is not obvious.

Tromp and Cilibrasi provide some empirical indications that minimum-length solutions for 1×1 Rush Hour may grow exponentially with puzzle size [164].

9.10   Triangular Rush Hour

Rush Hour has also inspired a triangular variant, called *Triagonal Slide-Out*, playable as an online applet [80]. The rules are the same as for Rush Hour (Section 9.9), except that the game is played on a triangular, rather than square, grid. Forty puzzles are available on the website.

Nondeterministic Constraint Logic gadgets showing Triagonal Slide-Out PSPACE-hard are shown in Figures 9.25 and 9.26. A and B cars represent inactive (outward-directed) input edges; C represents an inactive (inward-directed) output edge. We omit the proof of correctness.

One interesting feature of this triangular variant is that while it seems very difficult to build gadgets showing 1×1 Rush Hour hard, it might be much easier to show Triagonal Slide-Out with unit triangular cars hard, because for a unit triangular car to slide one unit, two triangular spaces must be empty: the one it is moving into, and the intervening space with the opposite parity.

Figure 9.26. How the gadgets are connected together.

9.11 Hinged Polygon Dissections

Properly speaking, this section is not about a game as defined in Section 1.1, because there is no concept of a discrete move: the problems are geometrical and continuous. Nonetheless, it is a further application of Nondeterministic Constraint Logic, and it demonstrates how continuous problems can sometimes be treated as if they were discrete.

We will simply define the problems and state the results here; we refer the reader to [86] (joint work with Greg Frederickson) for the detailed reductions.

A hinged dissection of a polygon is a dissection with a set of hinges connecting the pieces, so that they may kinematically reach a variety of configurations in the plane. Hinged polygon dissections have been a staple of recreational mathematics for at least the last century. One well known dissection is shown in Figure 9.27; this is Dudeney's [49] hinged dissection of a triangle to a square. There is an entire book [62] dedicated to hinged dissections.

The most basic problem is, given two configurations of a hinged polygon dissection, is it kinematically possible to go from one to the other? This problem and others are formalized as follows.

Terminology. We define a *piece* as an instance of a polygon. A *dissection* is a set of pieces. A *configuration* is an embedding of a dissection in the plane, such that only the boundaries of pieces may overlap. The *shape* of a

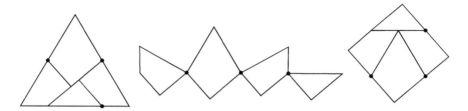

Figure 9.27. Dudeney's hinged triangle-to-square dissection.

configuration is the set of points, including the boundaries of pieces, that it occupies.

A *hinge point* of a piece is a given point on the boundary of the piece. A *hinge* for a set of pieces is a set of hinge points, one for each piece in the set. Given a dissection and a set of hinges, two pieces are *hinge-connected* if either they share a hinge or there is another piece in the set to which both are hinge-connected. A *hinging* of a dissection is a set of hinges such that all pieces in the dissection are hinge-connected.

A *hinged configuration* of a dissection and a hinging is a configuration that, for each hinge, collocates all hinge points of that hinge. A *kinematic solution* of a dissection, a hinging, and two hinged configurations is a continuous path from one configuration to the other through the space of hinged configurations. A *hinged dissection* of two polygons is a dissection and a hinging such that there are hinged configurations of the same shape as the polygons and there is a kinematic solution for the dissection, hinging, and hinged configurations. A hinged dissection of more than two polygons is similarly defined.

Decision questions. All of the hardness results are for problems of the form "is there a kinematic solution to..."; the difference is what we are given in each case. We are always given a dissection. We may or may not be given a hinging. We may be given two configurations, one configuration and one shape, or two shapes. The shapes may or may not be required to be convex. The pieces may or may not be required to be convex. We will always require that the configurations form polygonal shapes.

For each question, all of the information we are not given will form the desired answer. For example, if we are given a dissection, a hinging, and two shapes, the question is whether there exist satisfying configurations A and B forming the shapes, and a kinematic solution S from A to B.

The brief statement of our results is that if we are given a hinging, then all such questions are PSPACE-hard; if we are not given a hinging, but are given two configurations, then determining whether there is a hinging admitting a kinematic solution is PSPACE-hard.

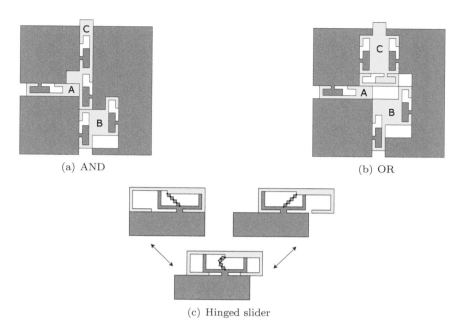

(a) AND   (b) OR

(c) Hinged slider

Figure 9.28.   NCL vertex gadgets for hinged dissections.

Gadgets. The Nondeterministic Constraint Logic gadgets required are shown in Figure 9.28. Essentially, these gadgets serve to turn a sliding-block puzzle into a hinged dissection: to go from one hinged configuration to another, one must effectively solve a sliding-block puzzle, thus an NCL problem. (See [86] for details.) Several other intricate gadgets are needed when the shapes are required to be convex.

Two-Player Games

In this chapter we present three new results for two-player games: Amazons, Konane, and Cross Purposes. Amazons is a relatively new game, about 20 years old, but it has received a considerable amount of study. Konane is a very old game, hundreds of years old at least; it has received some study, but not as much as Amazons. Cross Purposes is a brand new game.

10.1 Amazons

Amazons was invented by Walter Zamkauskas in 1988. Both human and computer opponents are available for Internet play, and there have been several tournaments, both for humans and for computers.

Amazons has several properties that make it interesting for theoretical study. Like Go, its endgames naturally separate into independent subgames; these have been studied using combinatorial game theory [10, 153]. Amazons has a very large number of moves available from a typical position, even more than in Go. This makes straightforward search algorithms impractical for computer play. As a result, computer programs need to incorporate more high-level knowledge of Amazons strategy [116, 126].

Playing an Amazons endgame optimally is NP-complete, as shown by Buro [20], leaving the complexity of the general game open.[1] We show

---

[1]Furtak, Kiyomi, Uno, and Buro independently showed Amazons to be PSPACE-complete at the same time as Hearn's result [69]. Curiously, [69] already contains two

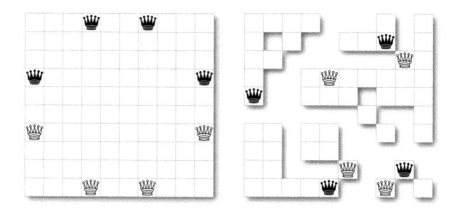

Figure 10.1.   Amazons start position and typical endgame position.

that generalized Amazons is PSPACE-complete [88, 92], by a reduction
from Bounded Two-Player Constraint Logic (2CL).

Amazons Rules.   Amazons is normally played on a 10×10 board.   The
standard starting position, and a typical endgame position, are shown in
Figure 10.1.   Each player has four *amazons*, which are immortal chess
queens. White plays first, and play alternates. On each turn a player must
first move an amazon, like a chess queen, and then fire an *arrow* from that
amazon. The arrow also moves like a chess queen. The square that the
arrow lands on is burned off the board; no amazon or arrow may move
onto or across a burned square. There is no capturing. The first player
who cannot move loses.

Amazons is a game of mobility and control, like Chess, and of territory,
like Go. The strategy involves constraining the mobility of the opponent's
amazons, and attempting to secure large isolated areas for one's own amazons. In the endgame shown in Figure 10.1, Black has access to 23 spaces,
and with proper play can make 23 moves; White can also make 23 moves.
Thus from this position, the player to move will lose.

---

different PSPACE-completeness proofs: one reduces from Hex, and the other from Generalized Geography. The paper is the result of the collaboration of two groups that
had also solved the problem independently then discovered each other. Thus, after
remaining an open problem for many years, the complexity of Amazons was solved
independently and virtually simultaneously by three different groups, using three completely different approaches, each of which leverages different aspects of the game to
construct gadgets.

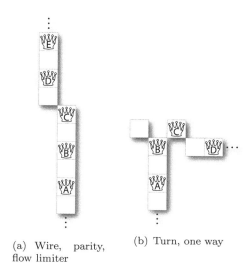

(a) Wire, parity,   (b) Turn, one way
flow limiter

Figure 10.2. Amazons wiring gadgets.

10.1.1   PSPACE-completeness

We reduce from Bounded 2CL, in Section 6.1.2. This requires construction of planar graphs made with AND, OR, FANOUT, CHOICE, and VARIABLE gadgets.

Basic Wiring. Signals propagate along *wires*, which will be necessary to connect the vertex gadets. Figure 10.2(a) shows the construction of a wire. Suppose that amazon A is able to move down one square and shoot down. This enables amazon B to likewise move down one and shoot down; C may now do the same. This is the basic method of signal propagation. When an amazon moves backward (in the direction of input, away from the direction of output) and shoots backward, we will say that it has *retreated*.

Figure 10.2(a) illustrates two additional useful features. After C retreats, D may retreat, freeing up E. The result is that the position of the wire has been shifted by one in the horizontal direction. Also, no matter how much space is freed up feeding into the wire, D and E may still only retreat one square, because D is forced to shoot into the space vacated by C.

Figure 10.2(b) shows how to turn corners. Suppose A, then B may retreat. Then C may retreat, shooting up and left; D may then retreat. This gadget also has another useful property: signals may only flow through it in one direction. Suppose D has moved and shot right. C may then move down and right, and shoot right. B may then move up and right, but it can

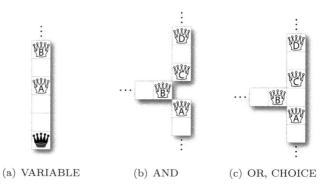

(a) VARIABLE   (b) AND   (c) OR, CHOICE

Figure 10.3. Amazons logic gadgets.

only shoot into the square it just vacated. Thus, **A** is not able to move up and shoot up.

By combining the horizontal parity shifting in Figure 10.2(a) with turns, we may direct a signal anywhere we wish. Using the unidirectional and flow-limiting properties of these gadgets, we can ensure that signals may never back up into outputs, and that inputs may never retreat more than a single space.

VARIABLE, AND, OR, CHOICE. The VARIABLE gadget is shown in Figure 10.3(a). If White moves first in a variable, he can move **A** down, and shoot down, allowing **B** to retreat later. If Black moves first, he can move up and shoot up, preventing **B** from ever retreating.

The AND and OR gadgets are shown in Figures 10.3(b) and 10.3(c). In each, **A** and **B** are the inputs, and **D** is the output. Note that, because the inputs are protected with flow limiters (Figure 10.2(a)), no input may retreat more than one square; otherwise the AND might incorrectly activate.

In an AND gadget, no amazon may usefully move until at least one input retreats. If **B** retreats, then a space is opened up, but **C** is unable to retreat there; similarly if just **A** retreats. But if both inputs retreat, then **C** may move down and left, and shoot down and right, allowing **D** to retreat.

Similarly, in an OR gadget, amazon **D** may retreat if and only if either **A** or **B** first retreats.

The existing OR gadget also suffices as a CHOICE gadget, if we reinterpret the bottom input as an output: if **B** retreats, then either **C** or **A**, but not both, may retreat.

FANOUT. Implementing a FANOUT in Amazons is a bit trickier. The gadget shown in Figure 10.4 accomplishes this. **A** is the input; **G** and **H** are the

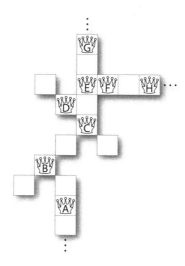

Figure 10.4. Amazons FANOUT gadget.

outputs. First, observe that until **A** retreats, there are no useful moves to be made. **C**, **D**, and **F** may not move without shooting back into the square they left. **A**, **B**, and **E** may move one unit and shoot two, but nothing is accomplished by this. But if **A** retreats, then the following sequence is enabled: **B** down and right, shoot down; **C** down and left two, shoot down and left; **D** up and left, shoot down and right three; **E** down two, shoot down and left; **F** down and left, shoot left. This frees up space for **G** and **H** to retreat, as required.

Winning. We will have an AND gadget whose output may be activated only if the white target edge in the 2CL game can be reversed; we need to arrange for White to win if he can activate this AND. We feed this output signal into a *victory* gadget, shown in Figure 10.5. There are two large rooms available. The sizes are equal, and such that if White can claim both of them, he will win, but if he can claim only one of them, then Black will win; we give Black an additional room with a single Amazon in it with enough moves to ensure this property.

If **B** moves before **A** has retreated, then it must shoot so as to block access to one room or the other; it may then enter and claim the accessible room. If **A** first retreats, then **B** may move up and left, and shoot down and right two, leaving the way clear to enter and claim the left room, then back out and enter and claim the right room.

Theorem 10.1. *Amazons is PSPACE-complete.*

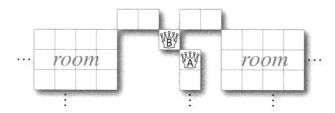

Figure 10.5. Amazons victory gadget.

Proof: Reduction from Bounded 2CL. Given a planar constraint graph made of AND, OR, FANOUT, CHOICE, and VARIABLE vertices, we construct a corresponding Amazons position, as described above. The reduction may be done in polynomial time: if there are k variables and l clauses, then there need be no more than $(kl)^2$ crossover gadgets to connect each variable to each clause it occurs in; all other aspects of the reduction are equally obviously polynomial.

As described, White can win the Amazons game if and only if he can win the corresponding 2CL game, so Amazons is PSPACE-hard. Since the game must end after a polynomial number of moves, it is possible to perform a search of all possible move sequences using polynomial space, thus determining the winner. Therefore, Amazons is also in PSPACE, and thus PSPACE-complete.   □

10.2   Konane

Konane is an ancient Hawaiian game, with a long history. Captain Cook documented the game in 1778, noting that at the time it was played on a 14×17 board. Other sizes were also used, ranging from 8×8 to 13×20. The game was usually played with pieces of basalt and coral, on stone boards with indentations to hold the pieces. King Kamehameha the Great was said to be an expert player; the game was also popular among all classes of Hawaiians.

More recently, Konane has been the subject of combinatorial game-theoretic analysis [21, 52]. Like Amazons, its endgames break into independent games whose values may be computed and summed. However, as of this writing, even $1 \times n$ Konane has not been completely solved, so it is no surprise that complicated positions can arise. We show the general problem to be PSPACE-complete.

Konane Rules. Konane is played on a rectangular board, which is initially filled with black and white stones in a checkerboard pattern. To begin the game, two adjacent stones in the middle of the board or in a corner are removed. Then, the players take turns making moves. Moves are made as in peg solitaire—indeed, Konane may be thought of as a kind of two-player peg solitaire. A player moves a stone of his color by jumping it over a horizontally or vertically adjacent stone of the opposite color, into an empty space. Stones so jumped are captured, and removed from play. A stone may make multiple successive jumps in a single move, as long as they are in a straight line; no turns are allowed within a single move. The first player unable to move wins.

10.2.1 PSPACE-completeness

The Konane reduction is similar to the Amazons reduction; the Konane gadgets are somewhat simpler. As before, the reduction is from Bounded Two-Player Constraint Logic (2CL). We need to build AND, OR, FANOUT, CHOICE, and VARIABLE gadgets.

Also as in the Amazons reduction, if White can win the constraint-logic game then he can reach a large supply of extra moves, enabling him to win. Black is supplied with enough extra moves of his own to win otherwise.

Basic Wiring. Wiring is needed to connect the vertex gadgets together. A Konane wire is simply a string of alternating black stones and empty spaces. By capturing the black stones, a white stone traverses the wire. Note that in the Amazons reduction, signals propagate by Amazons moving backward; in Konane, signals propagate by stones moving forward, capturing opposing stones.

Turns are enabled by adjoining wires as shown in Figure 10.6(a); at the end of one wire, the white stone comes to rest at the beginning of another, protected from capture by being interposed between two black stones. If the white stone tried to traverse the turn in the other direction, it would not be so protected, and Black could capture it. Thus, as in the Amazons reduction, the turn is also a one-way device, and we assume that gadget entrances and exits are protected by turns to ensure that signals can only flow in the proper directions.

Conditional Gadget. A single gadget serves the purpose of AND, FANOUT, and positional parity adjustment. It has two input/output pathways, with the property that the second one may only be used if the first one has already been used. This *conditional gadget* is shown in Figure 10.6(b); the individual uses are outlined below.

Observe that a white stone arriving at input 1 may only leave via output 1, and likewise for input 2 and output 2. However, if White attempts

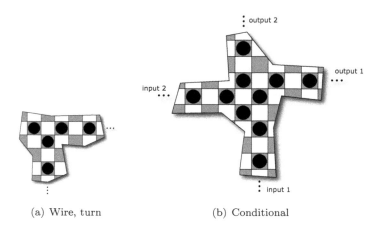

(a) Wire, turn   (b) Conditional

Figure 10.6. Konane wiring gadgets.

to use pathway 2 before pathway 1 has been used, Black can capture him in the middle of the turn. But if pathway 1 has been used, the stone Black needs to make this capture is no longer there, and pathway 2 opens up.

FANOUT, Parity. If we place a white stone within the wire feeding input 2 of a conditional gadget, then both outputs may activate if input 1 activates. This splits the signal arriving at input 1.

If we don't use output 1, then this FANOUT configuration also serves to propagate a signal from input 1 to output 2, with altered positional parity. This enables us to match signal parities as needed at the gadget inputs and outputs.

VARIABLE, AND, OR, CHOICE. The VARIABLE gadget consists of a white stone at the end of a wire, as in Figure 10.7(a). If White moves first in a variable, he can traverse the wire, landing safely at an adjoining turn. If Black moves first, he can capture the white stone and prevent White from ever traversing the wire.

The AND gadget is a conditional gadget with output 1 unused. By the properties of the conditional gadget, a white stone may exit output 2 only if white stones have arrived at both inputs. The OR gadget is shown in Figure 10.7(b). The inputs are on the bottom and left; the output is on the top. Clearly, a white stone arriving via either input may leave via the output.

As was the case with Amazons, the OR gadget also suffices to implement CHOICE, if we relabel the bottom input as an output: a white stone arriving along the left input may exit via either the top or the bottom.

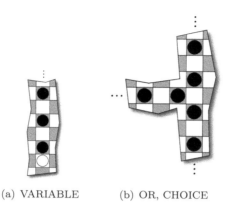

(a) VARIABLE   (b) OR, CHOICE

Figure 10.7. Konane VARIABLE, OR, and CHOICE gadgets.

Winning. We will have an AND gadget whose output may be activated just when White can win the given constraint-logic game. We feed this signal into a long series of turns, providing White with enough extra moves to win if he can reach them. Black is provided with his own series of turns, made of white wires, with a single black stone protected at the end of one of them, enabling Black to win if White cannot activate the final AND.

Theorem 10.2. *Konane is PSPACE-complete.*

Proof: Reduction from Bounded 2CL. Given a planar constraint graph made of AND, OR, FANOUT, CHOICE, and VARIABLE vertices, we construct a corresponding Konane position, as described above. As in the Amazons construction, the reduction is clearly polynomial. Also as in Amazons, White may reach his supply of extra moves just when he can win the constraint-logic game.

Therefore, a player may win the Konane game if and only if he may win the corresponding constraint-logic game, and Konane is PSPACE-hard. As before, Konane is clearly also in PSPACE, and therefore PSPACE-complete. □

10.3   Cross Purposes

Cross Purposes was invented by Michael Albert, and named by Richard Guy, at the Games at Dalhousie III workshop, in 2004. It was introduced to the authors by Michael Albert at the 2005 BIRS Combinatorial Game Theory Workshop. Cross Purposes is a kind of two-player version of the popular puzzle TipOver, which is NP-complete (Section 9.1; [90]). From

Figure 10.8. An initial Cross Purposes configuration, and two moves.

the perspective of combinatorial game theory [8, 27], in which game positions have values that are a generalization of numbers, Cross Purposes is fascinating because its positions can easily represent many interesting combinatorial game values.

Cross Purposes Rules. Cross Purposes is played on the intersections of a Go board, with black and white stones. In the initial configuration, there are some black stones already on the board. A move consists of replacing a black stone with a pair of white stones, placed in a row either directly above, below, to the left, or to the right of the black stone; the spaces so occupied must be vacant for the move to be made. See Figure 10.8. The idea is that a stack of crates, represented by a black stone, has been tipped over to lie flat. Using this idea, we describe a move as *tipping* a black stone in a given direction.

The players are called *Vertical* and *Horizontal*. Vertical moves first, and play alternates. Vertical may only move vertically, up or down; Horizontal may only move horizontally, left or right. All the black stones are available to each player to be tipped, subject to the availability of empty space. The first player unable to move loses.

We give a reduction from planar Bounded Two-Player Constraint Logic showing that Cross Purposes is PSPACE-complete.

10.3.1   PSPACE-completeness

The Cross Purposes construction largely follows those used for Amazons and Konane. To reduce from Bounded Two-Player Constraint Logic, we need AND, OR, FANOUT, CHOICE, and VARIABLE gadgets, and a way to wire them together into arbitrary graphs.

One new challenge in constructing the gadgets is that each player may only directly move either horizontally or vertically, but not both. Yet, for formula-game gadgets to work, one player must be able to direct signals two dimensionally. We solve this problem by restricting the moves of Horizontal so that, after the variable selection phase, his possible moves are constrained so as to force him to cooperate in Vertical's signal propaga-

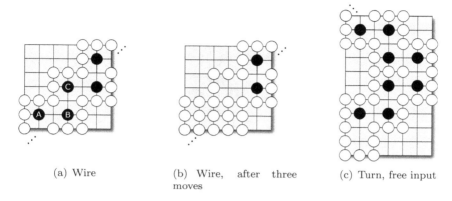

(a) Wire

(b) Wire, after three moves

(c) Turn, free input

Figure 10.9. Cross Purposes wiring.

tion. (We assume that the number of variables is even, so that it will be Vertical's move after the variable selection phase.) An additional challenge is that a single move can only empty a single square, enabling at most one more move to be made, so it is not obviously possible to split a signal. Again, we use the interaction of the two players to solve this problem.

We do not need a supply of extra moves at the end, as used for Amazons and Konane; instead, if Vertical can win the formula game, and correspondingly activate the final AND gadget, then Horizontal will have no move available, and lose. Otherwise, Vertical will run out of moves first, and lose.

Basic Wiring. We need wiring gadgets to connect the vertex gadgets together into arbitrary graphs. Signals flow diagonally, within surrounding corridors of white stones. A *wire* is shown in Figure 10.9(a). Suppose that Vertical tips stone **A** down, and suppose that Horizontal has no other moves available on the board. Then his only move is to tip **B** left. This then enables Vertical to tip **C** down. The result of this sequence is shown in Figure 10.9(b).

The turn gadget is shown in Figure 10.9(c); its operation is self-evident. Also shown in Figure 10.9(c) is a *free input* for Vertical: he may begin to activate this wire at any time. We will need free inputs in a couple of later gadgets.

Conditional Gadget. As with Konane (Section 10.2), a single *conditional gadget*, shown in Figure 10.10, serves the role of FANOUT, parity adjustment, and AND. A signal arriving along input 1 may only leave via output 1, and likewise for input 2 and output 2; these pathways are ordinary turns

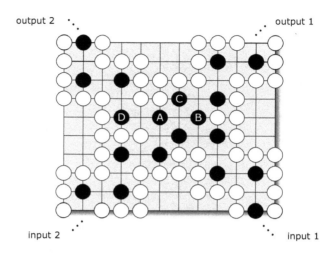

Figure 10.10. Cross Purposes conditional gadget.

embedded in the larger gadget. However, if Vertical attempts to use pathway 2 before pathway 1 has been used, then after he tips stone A down, Horizontal can tip stone B left, and Vertical will then have no local move. But if pathway 1 has already been used, stone B is blocked from this move by the white stones left behind by tipping C down, and Horizontal has no choice but to tip stone D right, allowing Vertical to continue propagating the signal along pathway 2.

FANOUT, Parity, AND. As with Konane, if we give Vertical a free input to the wire feeding input 2 of a conditional gadget, then both outputs may activate if input 1 activates. This splits the signal arriving at input 1.

If we don't use output 1, then this FANOUT configuration also serves to propagate a signal from input 1 to output 2, with altered positional parity. This enables us to match signal parities as needed at the gadget inputs and outputs. We must be careful with not using outputs, since we need to ensure that Vertical has no free moves anywhere in the construction; unlike in the constructions for Amazons and Konane, in Cross Purposes, there is no extra pool of moves at the end, and every available move within the layout counts. However, blocking an output is easy to arrange; we just terminate the wire so that Horizontal has the last move in it. Then Vertical gains nothing by using that output.

The AND gadget is a conditional gadget with output 1 unused. By the properties of the conditional gadget, output 2 may activate only if both inputs have activated.

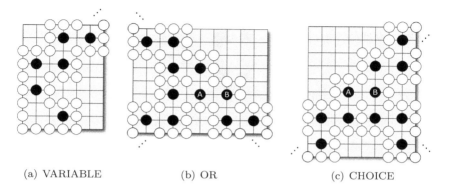

(a) VARIABLE   (b) OR   (c) CHOICE

Figure 10.11. Cross Purposes VARIABLE, OR, and CHOICE gadgets.

VARIABLE, OR, CHOICE. The VARIABLE gadget is shown in Figure 10.11(a). If Vertical moves first in a variable, he can begin to propagate a signal along the output wire. If Horizontal moves first, he will tip the bottom stone to block Vertical from activating the signal.

The OR gadget is shown in Figure 10.11(b). The inputs are on the bottom; the output is on the top. Whether Vertical activates the left or the right input, Horizontal will be forced to tip stone **A** either left or right, allowing Vertical to activate the output. Here we must again be careful with available moves. Suppose Vertical has activated the left input, and the output, of an OR. Now what happens if he later activates the right input? After he tips stone **B** down, Horizontal will have no move; he will already have tipped stone **A** left. This would give Vertical the last move even if he were unable to activate the final AND gadget; therefore, we must prevent this from happening. We will show how to do so after describing the CHOICE gadget.

As with Amazons and Konane, the existing OR gadget suffices to implement CHOICE, if we reinterpret it. This time the gadget must be rotated. The rotated version is shown in Figure 10.11(c). The input is on the left, and the outputs are on the right. When Vertical activates the input, and tips stone **A** down, Horizontal must tip stone **B** left. Vertical may then choose to propagate the signal to either the top or the bottom output; either choice blocks the other.

Protecting the OR Inputs. As mentioned above, we must ensure that only one input of an OR is ever able to activate, to prevent giving Vertical extra moves. We do so with the graph shown in Figure 10.12. Vertical is given a free input to a CHOICE gadget, whose output combines with one of the two OR input signals in an AND gadget. Since only one CHOICE

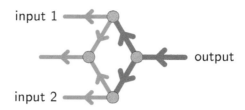

Figure 10.12. Protected OR.

output can activate, only one AND output, and thus one OR input, can activate. Inspection of the relevant gadgets shows that Vertical has no extra moves in this construction; for every move he can make, Horizontal has a response. (This construction is analogous to the protected OR used in Nondeterministic Constraint Logic (Section 5.2.3).)

Winning. We will have an AND gadget whose output may be activated only if the White player can win the corresponding constraint-logic game. We terminate its output wire with Vertical having the final move. If he can reach this output, Horizontal will have no moves left, and lose. If he cannot, then since Horizontal has a move in reply to every Vertical move within all of the gadgets, Vertical will eventually run out of moves, and lose.

Theorem 10.3. *Cross Purposes is PSPACE-complete.*

Proof: Reduction from Bounded 2CL. Given a planar constraint graph made of AND, OR, FANOUT, CHOICE, and VARIABLE vertices, we construct a corresponding Cross Purposes position, as described above. The reduction is clearly polynomial. Vertical may activate a particular AND output, and thus gain the last move, just when he can win the constraint-logic game.

Therefore, Cross Purposes is PSPACE-hard. As with Amazons and Konane, Cross Purposes is clearly also in PSPACE, and therefore PSPACE-complete. □

Perspectives on Part II

In Part II we have shown very many games and puzzles hard. Some of the proofs were difficult, especially as the proof technique was being developed, but some were very easy, once the proof technique was in place. For example, it took about half an hour to show Konane PSPACE-complete. Yet, in spite of a fair amount of study by combinatorial game theorists, and an important cultural history, no prior complexity results about Konane were known.

It is this kind of result that demonstrates the utility of constraint logic. A very large part of the work of reductions has already been done, and often one can simply select the kind of constraint logic appropriate for the problem at hand, and the gadgets will almost make themselves.

Most individual game complexity results are not particularly important. There are no game results in this book that are surprising, except for the undecidable version of constraint logic, and that is an abstract game. But taken as a whole, they demonstrate that the essential nature of games is captured effectively by the notion of constraint logic. Furthermore, they lend credence to the idea that an interesting game is almost always as hard as it "can" be: a bounded two-player game (without any trivial simplifying structure) ought to be PSPACE-complete, for example. Every additional result adds weight to this hypothesis.

Conclusions

In this section we summarize the contributions made by this book, and sketch some directions for future research.

12.1 Contributions

We have made four important contributions.

First, we have demonstrated a simple, uniform game framework, constraint logic, which concisely captures the concept of generalized combinatorial game. A constraint-logic game consists simply of a sequence of edge reversals in a directed graph, subject to simple constraints. There are natural versions of constraint logic for zero-, one-, two-player, and team games, both in bounded- and unbounded-length versions.

Second, we have demonstrated that each of these kinds of constraint-logic game corresponds to a distinct complexity class, or equivalently, to a distinct kind of resource-bounded computation, ranging from P-complete bounded, zero-player games, through PSPACE-complete unbounded puzzles, and up to undecidable (RE-complete) team games. This correspondence is proven by means of eight distinct reductions from Boolean formula games complete for the appropriate class to the corresponding constraint-logic game. For the undecidable team games, we also demonstrated that the existing Boolean formula game in the literature was in fact decidable, and independently derived a formula game that is actually undecidable.

Third, we have shown that the constraint-logic game framework makes hardness proofs for actual games and puzzles significantly easier. We have provided very many new proofs, mostly for problems that were either known to be open or previously unaddressed. In a few cases we rederived much simpler hardness proofs than the ones in the literature for games already known to be hard, thus explicitly demonstrating how much more concise and compact reductions from constraint logic can be, compared to conventional techniques (such as reducing directly from Satisfiability, Quantified Boolean Formulas, etc.). One key feature of constraint-logic games that often makes such reductions straightforward is that the hardness results for constraint logic apply even when the graphs are planar, across the spectrum of constraint-logic games (with the single exception of bounded, zero-player constraint logic). This means that there is no need to build "crossover" gadgets—often the most difficult component—in the actual game and puzzle reductions.

Finally, we have made more manifest the deep connection between the notions of game and of computation. Games are a natural generalization of conventional, deterministic computation. But a key difference from ordinary computation is that in a (generalized combinatorial) game, one is always dealing with a finite spatial resource. Any generalized combinatorial game can actually be played, physically, in the real world. But Turing machines, by contrast, are only idealized computers. We can never build a real Turing machine, because we cannot make an infinite tape.

The linchpin in this argument for games as computation is the undecidability result for team games with private information. Perfect play in such games is in direct correspondence with arbitrary computation on a Turing machine with an infinite tape. Yet, there are only a finite number of positions in the game. Thus, games represent a fundamentally distinct kind of computation.

12.2   Future Work

One direction for future work is obviously to apply the results here to additional games and puzzles, to show them hard. We list some candidates in Section A.11. Some of those games may yield to an application of constraint logic; others may not.

It is the ones that will not that are ultimately more interesting, such as 1×1 Rush Hour, and Subway Shuffle, which seem to lie right on the border between easy and hard problems. By attempting to hit them with the hammer of constraint logic, and observing how they fail to break, more can be learned about the mathematical nature of games.

We can also consider expanding our notion of a game. A very large unexplored space, from a constraint-logic perspective, is games involving random elements. A good starting point for developing constraint logic in this direction would be the probabilistic game automata of [26]. We could also consider games involving quantum information; research in this field is still at an early stage. On the upper end of the complexity/computability scale, now that we have reached undecidable games, is there anywhere left to go? We think so. For one thing, there are higher degrees of undecidability. Our undecidable games are RE-complete, corresponding to the power of a Turing machine with an infinite tape. What if we play a game on an infinite board? Could such a game perhaps correspond to a *hypercomputation* [30], a "computation" beyond the capacity of Turing machines?

But there is still interesting space to explore in the relatively more pedestrian classes of PSPACE, EXPTIME, EXPSPACE, etc. In particular, it is a bit unsatisfying that one needs to add the concepts of teams and of private information to move beyond ordinary two-player games. Isn't there some way to just add another kind of player? In general, an extra player represents an extra source of nondeterminism, and computational power. The notion of private information is merely another way of introducing an extra source of nondeterminism. In fact, one can explicitly add private information to a one-player game as well; it then becomes, effectively, a two-player game, because we might as well assume an adversary is choosing the private information. But what is the right way to look at private information in a two or more player game that makes it look like another player?

Similarly, the notion of disallowing repetitions in a game is using a kind of private information: the relevant history of the game is not present in the current configuration. Is there a way to translate that kind of hidden information, or nondeterminism, into another kind of player? It is questions such as these that continue to fascinate us.

Appendices

Survey of Games and
Their Complexities

This appendix attempts to serve as a concise reference to known complexity results for games and puzzles; it is modeled on Appendix A of Garey and Johnson's classic guide to NP-completeness [74]. For more background on the listed games and puzzles, see [36]. In general, we don't include games that are solvable in polynomial time, unless they are members of a family with hard variants.

Parts I and II of this book are organized by game type (number of players, bounded or unbounded), because the reduction techniques differ based on game type. Here, instead, we group games by family, to emphasize the similarities and differences between games in a family and to point out unexplored variants. In some cases one family can have games of different types. (For example, Section A.2.3, Block-Tipping and -Rolling Games, contains one-player bounded and unbounded puzzles and a two-player game.) However, our taxonomy is imperfect, as any such taxonomy must be; inevitably there will be games that can be naturally viewed as belonging to different families. (For example, the 15 Puzzle is clearly a sliding-block puzzle (Section A.2.1), but it could also be viewed as a game of sliding tokens on graphs (Section A.3.1), where the graph happens to be a square lattice.)

For most games that have concise definitions, we try to give formal problem statements. For others, or for games that exist in several different varieties, we will sometimes refer the reader to other sources for formal definitions.

One note on terminology: throughout, we use *horizontal* and *vertical* to refer to directions in the plane (usually thought of as a table surface, or a computer screen), parallel to the *x*- and *y*-axes. Occasionally we need to refer to directions out of the plane, in the *z*-direction: for example, we could have blocks standing on end. To avoid confusion, in such cases we are careful never to use the word *vertical.*

Some two-player games are described as either *impartial* or *partizan.* In impartial games the two players have identical moves available from any position; in partizan games the available moves are generally different.

A.1   Cellular Automata

A cellular automaton consists of a discrete set of *cells*, each of which can be in a finite number of *states*, and an *update rule* specifying the new states of all the cells as a function of the current states. Generally, the cells are arranged in a regular grid, and the update rule is applied independently at each cell as a function of a small set of nearby cells.

The most natural decision question to ask about a cellular automaton is: if the update rule is repeatedly applied from a given initial state, will a specified cell ever enter a specified state? There are many examples of "Turing universal" cellular automata for which this question is undecidable on an infinite grid; see, e.g., [173]. When restricted to a finite grid, the corresponding property is PSPACE-completeness (however, PSPACE-completeness may not hold if the computation is not encoded in a space-efficient manner). Cellular automata are generally not studied from this perspective; however, one cellular automaton for which PSPACE-completeness does hold is Conway's famous Game of Life:

> **LIFE**
>
> **Instance:** Rectangular grid of cells that are each *alive* or *dead.*
>
> **Question:** Will a given cell ever become alive, given the following update rule? On each step, a cell that is alive will remain alive if it has either two or three neighbors (horizontally, vertically, or diagonally) that are alive, otherwise it will die. A cell that is dead will become alive if it has exactly three neighbors that are alive.

Complexity: PSPACE-complete [141]. (Also see [8, 170].) (PSPACE-completeness is not mentioned explicitly in the cited works, but it does follow directly, at least from [141].)

Puzzle	Block Size	Constraints	Goal	Complexity
Warehouseman's Problem	arbitrary		configuration	PSPACE-complete
Sliding Blocks	fixed		location	PSPACE-complete
Rush Hour	fixed	H/V	location	PSPACE-complete
1 × 1 Rush Hour	1 × 1	H/V	location	?
15 Puzzle	1 × 1		configuration	polynomial
Lunar Lockout	1 × 1	slippery	location	NP-hard
Atomix	1 × 1	slippery, some immovable	partial configuration	PSPACE-complete

Table A.1. Summary of sliding-block puzzles.

A.2 Games of Block Manipulation

A large variety of games and puzzles are often framed as manipulations of rectangular blocks moving somehow in a grid or box. We divide these into *sliding-block puzzles*, *block-pushing puzzles*, and *block-tipping and -rolling games*.

A.2.1 Sliding-Block Puzzles

In a sliding-block puzzle, (usually) rectangular blocks are placed in a (usually) rectangular two-dimensional box; moves are made by sliding the blocks inside the box without overlapping. In the problem statements below, we assume unless otherwise stated that the blocks are rectangles of integral edge sizes and are placed on a grid, and that a move is an integral horizontal or vertical slide of a block. (Generally, these restrictions do not matter for the reductions.) Also, all blocks are distinct (labeled), though this does not always matter.

These puzzles can be characterized on a number of dimensions. Broadly, these are block size, movement constraints, and goal condition. We summarize the characteristics of the puzzles we consider in Table A.1: "H/V" means that some blocks may only slide horizontally and others vertically. "Slippery" means that when a block slides it must slide as far as it can in that direction. The goal is either to achieve a target total configuration or to move a particular block to a particular location. For many of the "location" results it is also hard to move the target block at all.

All games that we are aware of in this category are one-player unbounded games. One can imagine mutliplayer or bounded variants as well; these remain unexplored from a complexity standpoint. Additionally, some feature combinations suggest themselves as unexplored variants: specifically, (1 × 1, H/V, configuration) and (fixed, slippery, location) could be easier to show PSPACE-complete than 1 × 1 Rush Hour and Lunar Lockout, respectively.

WAREHOUSEMAN'S PROBLEM

Instance: Sliding-block puzzle.

Question: Is there a sequence of moves that eventually reaches a target configuration?

Complexity: PSPACE-complete [103]; earlier shown NP-hard [154]. The reduction requires blocks of arbitrary size. [85, Section 9.4] shows that the problem remains PSPACE-complete if the blocks are all 1×2.

SLIDING BLOCKS

Instance: Sliding-block puzzle with fixed-size blocks.

Question: Is there a sequence of moves that eventually moves a given block to a given location?

Complexity: PSPACE-complete, even when the blocks are all 1×2 and the goal is to move a given block at all [85, Section 9.3].

RUSH HOUR

Instance: Sliding-block puzzle with blocks of size 1×2 or 1×3. Horizontal blocks may only move horizontally; vertical blocks may only move vertically.

Question: Is there a sequence of moves that eventually moves a given block to a given location?

Complexity: PSPACE-complete, even when the goal is to move a given block at all [56, Section 9.9]. Also remains PSPACE-complete when the blocks are all 1×2 [164, 165]. If the blocks are 1×1 but still partitioned into horizontally and vertically constrained blocks, then the problem seems quite interesting; the complexity is open (but moving a given block at all is in P) [85, 164]. A triangular variant, Triagonal Slide-Out, is also PSPACE-complete (Section 9.9).

15 PUZZLE

Instance: Sliding-block puzzle with 1×1 blocks.

Question: Is there a sequence of moves that eventually reaches a target configuration?

Complexity: Polynomial [160]. Finding a solution using the minimum number of moves is NP-complete [135].

LUNAR LOCKOUT

Instance: Sliding-block puzzle with 1×1 blocks. Blocks must slide until they hit another block. (Sliding a block into the box edge is not allowed.)

Question: Is there a sequence of moves that eventually moves a given block to a given location?

Complexity: NP-hard [97]. Like 1×1 Rush Hour, it is an interesting open problem whether Lunar Lockout is PSPACE-complete. A variant allowing nonmovable blocks is PSPACE-complete [82].

ATOMIX

Instance: Sliding-block puzzle with 1×1 blocks; some blocks are immovable (*walls*). Movable blocks (*atoms*) are labeled with a type. Blocks must slide until they hit another block.

Question: Is there a sequence of moves that eventually assembles a given pattern of atom types (a *molecule*)?

Complexity: PSPACE-complete [100].

A.2.2 Block-Pushing Puzzles

These puzzles are similar to the sliding-block puzzles of Section A.2.1, but here there is a distinguished block, often called the *robot*, which is the only block that can be moved directly. The robot can, however, push other blocks. As with sliding-block puzzles, we generally assume that the blocks are all placed on a grid, and that they only move in integral units. Furthermore, the blocks are all generally 1×1.

The puzzles vary based on what happens when the robot attempts to move into a block. First, the robot may push up to some number k of blocks, with an empty space at the far end; a move is blocked if there are more than k blocks in a row. Fixed (immovable) blocks may or may not be allowed. When a block is pushed, the "physics" may dictate different results: in the basic version, the robot and the pushed blocks all move one unit in the appropriate direction. In the PushPush variants, the robot

Puzzle	Blocks	Fixed	Motion	Revisit	Goal	Complexity
Sokoban	1	yes	push	yes	configuration	PSPACE-complete [33, 85, Section 9.7]
Push-k	k	no	push	yes	location	NP-hard [37]
Push-2-F	2	yes	push	yes	location	PSPACE-complete [40]
Push-k-X	k	no	push	no	location	NP-complete [37]
PushPush-k	k	no	slide	yes	location	PSPACE-complete [42]
PushPush-k-X	k	no	slide	no	location	NP-complete [37]
Push-*	∞	no	push	yes	location	NP-hard [98]
Push-*-F	∞	yes	push	yes	location	PSPACE-complete [15,40]
Push-*-X	∞	no	push	no	location	NP-complete [37, 98]
PushPush-*	∞	no	slide	yes	location	NP-hard [42, 98]
PushPush-*-X	∞	no	slide	no	location	NP-complete [37, 98]
Push-1-G	1	no	gravity	yes	location	NP-hard [67]

Table A.2. Summary of block-pushing puzzles.

moves one unit, but the pushed blocks slide until they hit another block. One version also considers gravity acting on blocks after a push. The goal of the puzzle is either to reach a specified configuration or to move the robot to a given location. Finally, there are variants where the robot is not allowed to revisit its own path. These puzzles are largely defined by this set of properties, listed in Table A.2; we refer the reader to the given references for details.

The study of this type of puzzle was initiated by Wilfong [171], who proved that deciding whether the robot can reach a desired target is NP-hard when the robot can push and pull L-shaped blocks (not listed in table). When the final position of the blocks matters, the problem becomes PSPACE-complete.

A.2.3   Block-Tipping and -Rolling Games

In these games the three-dimensional nature of the blocks becomes evident: instead of being slid or pushed, as in Sections A.2.1 and A.2.2, the blocks must be moved around the board by tipping them along an edge. Here we assume that the blocks are cuboids with integral edge lengths k, m, n, and that they occupy grid-aligned positions on the board. For a move to be legal, there must be vacant space within the grid to accommodate the

Game	Players	Length	Goal	Complexity
TipOver	1	bounded	see problem	NP-complete
Rolling Block Maze	1	unbounded	location	PSPACE-complete
Cross Purposes	2	bounded	last move	PSPACE-complete

Table A.3. Summary of block-tipping and -rolling games.

block face tipping into the grid. We list three such games in Table A.3 and define them below. A rolling-cube puzzle that has a more complicated definition, requiring face labels to match some grid labels, is shown to be NP-complete in [19].

TIPOVER

Instance: Set of $1 \times 1 \times n$ blocks initially placed standing on end in a square grid, *tipper* atop one of the blocks, target block.

Question: Can the tipper reach the target block? Moves are made by tipping over the block the tipper is standing on, or by moving the tipper to any block touching the block it is standing on. Once a block has tipped over it may no longer move.

Complexity: PSPACE-complete, even if the blocks are all $1 \times 1 \times 2$ [90, Section 9.1].

ROLLING BLOCK MAZE

Instance: Set of blocks in a square grid, target block, target placement for this block.

Question: Is there a sequence of moves that places the target block in its target placement?

Complexity: PSPACE-complete, even if the blocks are all $1 \times 1 \times 2$ or $1 \times 1 \times 3$ [18]. The question of the complexity with only $1 \times 1 \times 2$ blocks is open. Note that in "traditional" rolling-block mazes there is a single block that may roll; such mazes are solvable in linear time. (In [18], there are also forbidden squares, but these may be replaced by vertical 1×3 blocks. If forbidden squares are used, all the blocks may be $1 \times 1 \times 2$.)

CROSS PURPOSES

Instance: Set of $1 \times 1 \times 2$ blocks in a square grid.

Question: Can Player H win the following game? H and V take turns tipping blocks that are standing on a 1×1 face. H can only tip blocks horizontally; V can only tip blocks vertically. The last player to move wins.

Complexity: PSPACE-complete [92, Section 10.3]. The problem as formulated in Section 10.3 is played with Go stones on a Go board, but here we emphasize the similarity to other block-rolling games. (Also, TipOver was an inspiration for the game design.)

A.3 Games of Tokens on Graphs

A large number of games involve the placement of tokens on the vertices of a graph and/or the movement of tokens along graph edges.

A.3.1 Moving Tokens

In these games tokens move from vertex to vertex along the edges of a (generally) directed graph. The games vary based on matching requirements: there can be a distinct set of edges each type of token is allowed to move on (token-edge), a distinct type of token each player may move (player-token), or a distinct set of edges each player may move tokens on (player-edge). The games also vary by what happens when a destination vertex is already occupied: either or both tokens may be removed, or the move may be blocked. Finally, the games vary based on the goal: last move, capturing a token, or moving a particular token to a specified target. There are also games that are similar, but have additional constraints on adjacency of tokens; these are addressed in Section A.3.2.

We summarize several pursuit games in Table A.4. Many of the games have separate results for general digraphs, where the games are unbounded, and for directed acyclic graphs (dags), where the games are bounded. Many of these games are fully defined by the characteristics we have listed; we omit formal problem statements. See the cited references for details. We do add the following notes:

- For *Capture*, a player may move a token to an occupied vertex only if it contains an opponent's token, in which case that token is captured (removed), as in *Hit*.

- In *Pursuit*, or *Cops and Robbers*, one player (the *robber*) has a single token; the other player (the *cops*) has multiple tokens and moves all his tokens on each turn. The cops win if a cop token ever moves to a vertex occupied by the robber.

- *Subway Shuffle* empirically seems hard, but does not seem amenable to a reduction from either NCL or any other standard problem.

Game	Players	Matching	A → B	Goal	Complexity
Annihilation	2	token-edge	A, B removed	last move	PSPACE-hard [57]
...on a dag					PSPACE-complete [57]
Remove	2	token-edge	A removed	last move	NP-hard [59]
...on a dag					NP-hard [59]
Hit	2	token-edge	B removed	last move	PSPACE-hard [57]
...on a dag					PSPACE-complete [57]
Capture	2	player-token	see text	last move	EXPTIME-complete [78]
...on a dag					PSPACE-complete [78]
Contrajunctive	2	player-edge	A, B removed	last move	NP-hard [59]
...on a dag					NP-hard [59]
Node Blocking	2	player-token	move blocked	last move	EXPTIME-complete [78]
...on a dag					PSPACE-complete [46]
Edge Blocking	2	player-edge	move blocked	last move	PSPACE-hard [57]
...on a dag					PSPACE-complete [57]
Pursuit	2	player-token	see text	capture	EXPTIME-complete [78]
...on a dag					PSPACE-complete [78]
Subway Shuffle	1	token-edge	move blocked	target	? [89,91]

Table A.4. Summary of pursuit games. "A → B" denotes token A moving to a vertex occupied by token B.

- *Target*, previously shown EXPTIME-complete [157] (not listed in table), is a weaker form of *Node Blocking* in which players can additionally win by moving one of their tokens to a vertex that is one of their targets. Target is PSPACE-complete for dags even when the graph is bipartite and only one player has targets [78]. There is also a semi-partizan variant of Target (see reference).

A.3.2   Games with Adjacency Constraints

In these games tokens are placed or moved on a graph with restrictions on when two tokens may be adjacent. These restrictions make many otherwise easy problems hard. The prototypical problem in this family is *Independent Set*, which can be viewed as a bounded one-player game. The other games are natural variants of Independent Set. We summarize several games with adjacency constraints in Table A.5.

Game	Players	Prohibited Adjacency	Goal	Complexity
Independent Set	1	any	place n	NP-complete
Sliding Tokens	1	any	move target	PSPACE-complete
Node Kayles	2	any	last move	PSPACE-complete
Partizan Node Kayles	2	any	last move	PSPACE-complete
Snort	2	different colors	last move	PSPACE-complete
Col	2	same colors	last move	?

Table A.5. Summary of graph games with adjacency constraints.

INDEPENDENT SET

Instance: Graph G; integer n.

Question: Can n tokens be placed on the vertices of G such that those vertices form an independent set (they do not share an edge)?

Complexity: NP-complete [74].

SLIDING TOKENS

Instance: Graph with tokens on some vertices, forming an independent set.

Question: Is there a sequence of moves that eventually moves a given token? A move consists of sliding a token along an edge to an adjacent vertex, again forming an independent set.

Complexity: PSPACE-complete [85, Section 9.5].

NODE KAYLES

Instance: Graph G.

Question: Can Player 1 win the following game? Players 1 and 2 alternately place tokens on vertices of G, such that all the vertices with tokens form an independent set. The first player unable to move loses.

Complexity: PSPACE-complete [147]. *Partizan Node Kayles* is identical except that G is partitioned into V_1 and V_2; Player i may only play on vertices from V_i. It is also PSPACE-complete [147].

SNORT

Instance: Graph G.

Question: Can Player 1 win the following game? Players 1 and 2 alternately place tokens on vertices in G, such that no token is ever adjacent to a token played by the other player. The first player unable to move loses.

Complexity: PSPACE-complete [147].

COL

Instance: Graph G.

Question: Can Player 1 win the following game? Players 1 and 2 alternately place tokens on vertices in G, such that each player's selected vertices form an independent set. The first player unable to move loses.

Complexity: Open [8]. This game is in a sense the opposite of Snort.

A.3.3 Generalized Geography

In *(Generalized) Geography*, two players alternately move a single token, and then erase part of the graph. The goal is to move last.

DIRECTED/UNDIRECTED EDGE/VERTEX GEOGRAPHY

Instance: (Directed, undirected) graph, token placed on an initial vertex.

Question: Can Player 1 win the following game? Players 1 and 2 take turns, where a turn consists of moving the token to an adjacent vertex, and then removing either the edge traversed (Edge Geography) or the vertex moved from (Vertex Geography) from the graph. The first player unable to move loses.

Complexity: See Table A.6.

Edges	Element Removed	Complexity
directed	edge	PSPACE-complete [147]
directed	vertex	PSPACE-complete [114]
undirected	edge	PSPACE-complete [60]
undirected	vertex	polynomial [60]

Table A.6. Summary of Geography variants.

Game	Players	Remove	Comments	Goal	Complexity
Peg Solitaire	1	yes		one peg	NP-complete [168]
Peg Duotaire	2	yes	impartial	last move	? [124]
HotSpot	1	no	see text	location	?
Konane	2	yes	partizan	last move	PSPACE-complete [91,92, Section 10.2]

Table A.7. Summary of peg-jumping games.

A.4 Peg-Jumping Games

Peg-jumping games are games played with pegs or other tokens on a grid, where a move consists of "jumping" a peg over adjacent pegs into an empty space. Depending on the game, moves may also cause the peg(s) jumped over to be removed from the board. These games can have one or two players and can be impartial or partizan. Table A.7 summarizes the rules for each game; we refer the reader to the listed references for full definitions. In *HotSpot*, tokens may be *small* or *large*. Two large tokens may not be adjacent. Empirically this seems to add complexity to the game, but the status is open. In *Peg Duotaire* and *Konane*, players may make multiple jumps in one turn. In Konane, such jumps must be in a straight line. Konane also has pieces of two different colors, and players can only move pieces of their color.

A.5 Connection Games

In connection games, players try to complete a path between two vertices by sequentially choosing a set of vertices or edges from a graph (often a grid with the sides connected to the target vertices). All the games we consider here are two-player partizan games. In Table A.8 we list general characteristics and complexities of these games, and refer the reader to the

Game	Choose	Comments	Complexity
Hex	vertices	diamond section of triangular graph	PSPACE-complete [140]
Shannon Switching Game on Edges	edges		polynomial [17]
Shannon Switching Game on Edges	edges	directed graph	PSPACE-complete [53]
Twixt	vertices	edges are knight's moves on a square grid graph	?, related problem NP-complete [119]

Table A.8. Connection games.

Game	Complexity
Othello (Reversi)	PSPACE-complete [106]
Checkers	EXPTIME-complete [145]
Chess	EXPTIME-complete [58]
Shogi	EXPTIME-complete [2, 177]
Go	EXPTIME-complete [143]
Amazons	PSPACE-complete [69, 92, Section 10.1]
Gomoku (Gobang)	PSPACE-complete [139]
Mastermind	NP-complete [161]
Phutball (Philosopher's Football)	PSPACE-hard [47]

Table A.9. Summary of classic board games.

listed references for full definitions. (Note that *Hex* is a special case of the *Shannon Switching Game on Vertices*, earlier shown PSPACE-complete [53].)

A.6 Other Board Games

For these classic board games, we merely state the complexities in Table A.9 and refer the reader to the indicated references or to [36] for their full definitions.

For *Go*, the result applies only to Japanese rules. In particular if the *superko* rule is used, the lower bound drops to PSPACE [114], and the upper bound rises to EXPSPACE. Some other decision questions about restricted Go positions are NP-complete [32] or PSPACE-complete [31, 172].

For *Mastermind*, one natural decision question is NP-complete; others remain open.

For *Phutball*, even determining whether there is a one-move win is NP-complete [38].

A.7 Pencil Puzzles

These puzzles, largely popularized by Japanese publisher Nikoli, involve filling in a grid in some manner (with numbers, lines, shading, etc.) to satisfy given local and global properties. The most well known of these puzzles is *Sudoku*. In general, these puzzles all tend to be NP-complete. (They cannot be any harder, because they are all bounded one-player games.) Some are also *ASP-complete* [175], which implies NP-completeness as well as NP-completeness of finding another solution when given one or more other solutions to the same problem. ("ASP" stands for "Another Solution Problem.")

In all of these problems, *adjacent* means horizontally or vertically adjacent, and *connected* means connected via horizontal and vertical connections.

SUDOKU

Instance: $n^2 \times n^2$ grid, divided into $n \times n$ subgrids, with numbers in some cells.

Question: Can the empty cells be filled in with the numbers $1 \ldots n^2$ such that each row, each column, and each subgrid contains all of $1 \ldots n^2$?

Complexity: NP-complete, ASP-complete [175, 176].

KAKURO (CROSS SUM)

Instance: A polyomino (a rectangular grid where only some squares may be used), and an integer for each maximal contiguous (horizontal or vertical) strip of squares.

Question: Can each square be filled with a digit between 1 and 9 such that each strip has the specified sum and has no repeated digit?

Complexity: NP-complete, ASP-complete [175].

LIGHT UP (AKARI)

Instance: Rectangular grid in which squares are either rooms or walls and some walls have a specified integer between 0 and 4.

Question: Can lights be placed in a subset of the rooms such that each numbered wall has exactly the specified number of adjacent lights, every room is horizontally or vertically visible from a light, and no two lights are horizontally or vertically visible from each other?

Complexity: NP-complete, ASP-complete [121, 122].

SLITHERLINK (FENCES)

Instance: Rectangular grid with labels between 0 and 4 in some cells.

Question: Is there a simple cycle on the edges of the cells such that each labeled cell is surrounded by the specified number of edges?

Complexity: NP-complete, ASP-complete [175, 176].

NONOGRAM (PAINT BY NUMBERS)

Instance: A rectangular grid with a sequence of integers on each row and column.

Question: Can a subset of the squares in the grid be filled such that in each row and column, the maximal contiguous runs of filled squares have lengths that match the specified sequence?

Complexity: NP-complete, ASP-complete [167].

FILLOMINO

Instance: A rectangular grid with some squares filled with positive integers.

Question: Can the remaining squares be filled with positive integers such that every maximal connected region of equally numbered squares consists of exactly that number of squares?

Complexity: NP-complete, ASP-complete [176].

LITS

Instance: A rectangle divided into polyomino pieces.

Question: Can a tetromino (connected subset of four squares) be placed in each polyomino such that the union of tetrominoes is connected, yet induces no 2 × 2 square, and no touching tetrominoes have the same shape?

Complexity: NP-complete, ASP-complete [122].

MASYU (PEARL PUZZLE)

Instance: Rectangular grid with some cells containing white or black pearls.

Question: Is there a simple path through the squares that visits every pearl, turns 90° at every black pearl, does not turn immediately before or after black pearls, goes straight through every white pearl, and turns 90° immediately before or after every white pearl?

Complexity: NP-complete [66].

HITORI

Instance: Rectangular grid with each square labeled with an integer.

Question: Is it possible to paint some squares such that every row and every column has no repeated unpainted label, painted squares are never adjacent, and unpainted squares are connected?

Complexity: NP-complete [94, Section 9.2].

NURIKABE

Instance: Rectangular grid with some squares labeled with positive integers.

Question: Is there a connected set of unlabeled squares that induces no 2×2 square, and whose removal results in exactly one region per labeled square, whose size equals that label?

Complexity: NP-complete [102, 120].

TENTAI SHOW (SPIRAL GALAXIES)

Instance: A rectangular grid with dots at some vertices, edge midpoints, and face centroids.

Question: Can the rectangle be divided into exactly one polyomino piece per dot, that is two-fold rotationally symmetric
around the dot?

Complexity: NP-complete [68].

BAG (CORRAL PUZZLE)

Instance: A rectangular grid with some squares labeled with
positive integers.

Question: Is there a simple cycle on the grid that encloses all
labels such that the number of squares horizontally and vertically visible from each labeled square equals the label?

Complexity: NP-complete [65].

HIROIMONO (GOISHI HIROI)

Instance: A rectangular grid with stones at some of the vertices.

Question: Is there a path that visits all stones, removing each
as it as visited, and turns only at the stones, optionally
changing direction by ±90°?

Complexity: NP-complete [4].

HEYAWAKE

Instance: A rectangular grid subdivided into rectangular rooms,
some of which are labeled with a positive integer.

Question: Is it possible to paint some squares such that the
number of painted squares in each labeled room equals the
label, painted squares are never adjacent, unpainted squares
are connected, and contiguous (horizontal or vertical) strips
of unpainted squares intersect at most two rooms?

Complexity: NP-complete [99].

MORPION SOLITAIRE

Instance: A configuration of points drawn at the intersections of a square grid, integer n.

Question: Is it possible to make n moves? A move consists of placing a new point at a grid intersection, and then drawing a horizontal, vertical, or diagonal line segment connecting k consecutive points that include the new one. Segments with the same direction may not overlap (disjoint model), or may overlap only at a common endpoint (touching model).

Complexity: NP-complete for either model, for $k > 4$ [43]. There are approximation results for $k = 4$.

Finally, we mention the pencil game *Battleships* or *Battleship Solitaire*, in which several ships must be placed into a partially filled grid, with specified numbers of ship segments occupying each row and column. This problem is NP-complete [148]; see the reference for the formal definition.

A.8   Formula Games

There is quite a large number of complexity results for problems stated in terms of Boolean formulas, many of which can be interpreted as games. Here we restrict our attention to canonical formula games for different game categories, and we refer the reader to the cited references or to [74] or [147] for more variants. For basic definitions related to Boolean formulas, see Section B.5.1.

SATISFIABILITY (SAT)

Instance: Boolean formula ϕ.

Question: Is there an assignment to the variables of ϕ such that ϕ is true?

Complexity: NP-complete [28], even if ϕ is in 3CNF (3SAT). Satisfiability can be seen as a bounded one-player game where the moves are to assign variables and the goal is to satisfy the formula.

QUANTIFIED BOOLEAN FORMULAS (QBF)

Instance: Quantified Boolean formula ϕ.

Question: Is ϕ true?

Complexity: PSPACE-complete [158], even if ϕ is in 3CNF. QBF can be seen as a bounded two-player game where the moves are to alternately assign variables, with one player trying to make ϕ true and the other false.

G_6

Instance: CNF Boolean formula F in variables $X \cup Y$, variable assignment α.

Question: Does Player I have a forced win in the following game? Players I and II take turns. Player I (II) moves by changing at most one variable in X (Y); passing is allowed. Player I wins if F ever becomes true.

Complexity: EXPTIME-complete [157]. Related games G_1 through G_5 are also EXPTIME-complete. These are all unbounded two-player games.

G_1'

Instance: 4CNF Boolean formula F in variables $X \cup Y \cup \{t\}$, variable assignment α.

Question: Does Player I have a forced win in the following game? Players I and II take turns. Player I moves by setting t to true and setting the variables in X to any values; Player II moves by setting t to false and setting the variables in Y to any values. A player loses if F is false after his move, or if the variable assignment after his move has the same value as after some earlier move.

Complexity: EXPSPACE-complete [144]. This is a "no-repeat" version of Stockmeyer and Chandra's game G_1 mentioned above. The corresponding no-repeat versions of G_2 and G_3 are also EXPSPACE-complete. A further rule modification, adding two special variables, results in game G_1^*, which is 2EXPTIME-complete.

A.9   Other Games

In this section we list various games that don't fit neatly into any other category. As the definitions are often not simple, for most games we don't provide formal problem statements, and we refer the reader to the cited references for details.

HACKENBUSH

Instance: Graph with each edge colored either red, blue, or green, and with some vertices marked as *rooted*.

Question: Can Red win the following game? Players Red and Blue take turns removing an edge of an appropriate color (either their own color or green), which also causes all edges not connected to a rooted vertex to be removed. The first player unable to move loses.

Complexity: NP-hard, even for graphs with no green edges [8, pp. 189–227].

STRINGS-AND-COINS

Instance: Graph with loop edges allowed.

Question: Can Player 1 win the following game? Players 1 and 2 take turns removing an edge from the graph ("cutting a string"). If a move leaves a vertex (a "coin") with no edges attached, the player takes another turn. The winner is the player who isolates the most vertices ("collects the most coins").

Complexity: NP-hard [8, pp. 577–578]. (In [8], strings may be connected to the ground; we model this here with loop edges.) The popular game *Dots-and-Boxes* [8, 9] is isomorphic to Strings-and-Coins with appropriate restrictions placed on the graph. Eppstein [50] observes that the Strings-and-Coins reduction should also apply to Dots-and-Boxes.

INSTANT INSANITY

Instance: n cubes, with each face colored one of n colors.

Question: Is it possible to stack the cubes so that each color appears exactly once on each of the four sides of the stack?

Complexity: NP-complete [142]. A two-player version is PSPACE-complete.

ALTERNATING MAXIMUM WEIGHTED MATCHING

Instance: Weighted graph G, integer bound B.

Question: Does Player 1 have a forced win in the following game? Players 1 and 2 alternate choosing a new edge from G, such that no edge can share an endpoint with any already chosen edge. If the sum of the weights of chosen edges ever exceeds B, Player 1 wins.

Complexity: PSPACE-complete [74, p. 256].

CROSSWORD PUZZLE CONSTRUCTION

Instance: List of words, rectangular grid with squares marked as either obstacles or blank.

Question: Can a subset of the words be placed into horizontally or vertically maximal blank strips so that crossing words have matching letters?

Complexity: NP-complete [74, p. 258], even when the grid has no obstacles so every row and column must form a word.

Tetris. In the popular computer game *Tetris*, tetrominoes fall from above and may be rotated and shifted as they fall. When they contact other pieces they stop falling, and any completely filled rows disappear, bringing any rows above down one level. If the sequence of pieces is given, it is NP-complete to determine whether it is possible to keep the stack below a finite height [16]. Other problems are NP-complete to approximate.

Dyson Telescope Game. The *Dyson Telescope Game* is an online puzzle produced by the Dyson corporation, based on their telescoping vacuum cleaners. The goal is to maneuver a ball on a square grid from a starting position to a goal position by extending and retracting telescopes on the grid, thus pushing or pulling the ball. The basic problem is PSPACE-complete; restricted versions (which are nonetheless interesting) are polynomial [44].

Mahjong Solitaire. *Mahjong Solitaire* or *Shanghai* is a computer game played with Mahjong tiles stacked in a pattern that hides some tiles. Each move removes a pair of matching tiles that are completely exposed. The goal is to remove all tiles. If all the tile positions are known, the problem is NP-complete [50]. An approximation problem is PSPACE-hard when the positions are unknown but uniformly distributed [25].

Plank Puzzle. A *Plank Puzzle* is a one-player unbounded game in which the moves are to walk across wooden planks, with the ability to pick up and carry one plank at a time and deposit it in another location where it will fit. The goal is to reach a particular location. Plank Puzzles are defined and shown PSPACE-complete in [87, Section 9.6].

Cryptarithms. *Cryptarithms* are puzzles such as SEND + MORE = MONEY, where the goal is to find an assignment of digits to letters that satisfies the equation. A generalization of this problem to bases other than decimal is NP-complete [51].

Minesweeper. *Minesweeper* is a well-known imperfect-information computer game with the goal of discovering the locations of a set of mines without setting one off. A move consists of uncovering a square. If that square contains a mine, the player loses; otherwise the number of mines in the 8 adjacent squares is placed in that square. Testing consistency of currently exposed squares is NP-complete [107]. From the point of view of this book, the "natural" decision question is whether the player has a forced win from a given position. This problem is coNP-hard; if the position is assumed consistent, then it is coNP-complete [91, 93].

Minesweeper can be seen as an "anti-puzzle." An ordinary puzzle, or one-player game, can be thought of as a degenerate two-player game, where one player's moves are forced: the player makes a move, something deterministic happens, then the player moves again, etc. In Minesweeper, it turns out that the order of the player's moves, so long as they are known safe based on the revealed state, is irrelevant: the question is whether the computer can place further mines from a given partial position so as to prevent a forced win. So, from a certain point of view, the player's moves are forced, and the game's moves are free, just the opposite of a normal puzzle.

Clickomania. *Clickomania* or *Same Game* is a computer puzzle consisting of a rectangular grid of colored square blocks. Adjacent blocks of the same color are considered part of the same group. A move selects a group containing at least two blocks and removes them; blocks then fall to fill any holes, and columns slide to remove any empty columns. The goal is to remove all the blocks. Deciding whether this is possible is NP-complete [11]. The case of only two colors remains open; a natural two-player variant is also open.

Vexed. *Vexed* or *Cubic* is a computer puzzle consisting of a rectangular grid with fixed blocks and movable colored blocks (all unit squares). A move consists of sliding a colored block left or right one unit, after which the block falls according to gravity. If the block is adjacent to any other blocks of the same color in its final position, then all such blocks disappear (as if clicking the group in Clickomania). The goal is to eliminate all colored blocks. Deciding whether this is possible is NP-complete [64].[1]

KPlumber. *KPlumber* is a computer puzzle consisting of a rectangular grid of square tiles, where each side of each tile is marked either "open" or "closed." A move consists of rotating a tile by 90° in place. The goal is to rotate the tiles so that every two neighboring tiles share matching sides, either both open or both closed. Deciding whether this is possible is NP-complete [110], even with specific tile sets: tiles with exactly zero or two open sides, or tiles with exactly two or three open sides.

Wriggle puzzles. *Wriggle puzzles* are a kind of sliding-piece puzzle, where the sliding pieces are flexible worms (*wrigglers*). A puzzle consists of a box with some internal barriers and some number of wrigglers; the goal is to slide the wrigglers backwards and forwards so that a particular wriggler can reach a certain location within the box. Wriggle puzzles were invented by Andrea Gilbert [76]. These puzzles are PSPACE-complete [118].

Klondike. *Klondike* or *Solitaire* is the classic solitaire card game. In the perfect information version of this game, we suppose the player knows all of the normally hidden cards. This game is NP-complete (when generalized to n cards per suit), even with just three suits [117]. Additional results in [117] are that Klondike with one black suit and one red suit is NL-hard, Klondike with any fixed number of black suits and no red suits is in NL, and Klondike with one suit is in AC^0, among other results.

FreeCell. *FreeCell* is another common solitaire card game. We will not attempt to describe the rules here. FreeCell is NP-complete (again, when generalized to n cards per suit), for any fixed positive number of free cells [95].

Jigsaw puzzles. *Jigsaw puzzles* can be formalized in a variety of ways, with different edge matching rules, different boundary conditions, and different constraints on numbers of available pieces. Many versions are NP-complete [74, p. 257] [35]. Infinite generalizations are undecidable [7, 35].

---

[1]David Eppstein [50] points out that [64] establishes only NP-hardness, while the problem is not obviously in NP. Friedman and Hearn together showed (in personal communication) that the problem is in NP as well.

Packing puzzles. In a *packing puzzle*, the goal is to fit all of a given set of shapes (often polyominoes) into a target shape without overlap. Many varieties of this basic type of puzzle are NP-complete [12, 35, 113, 125].

Reflections. In *Reflections*, we are given a rectangular grid with one square containing a laser, some squares containing light bulbs, some squares marked one-way in an axis-parallel direction, and the remaining squares marked either empty or wall. We are also given a number of diagonal mirrors and T-splitters that we can place into empty squares. The light then travels from the laser through the grid, following appropriate rules when meeting each board element. The goal is to place the mirrors and splitters so that each light bulb gets hit an odd number of times. This puzzle is NP-complete [108].

Reflexion. In *Reflexion* [101], we are given a rectangular grid in which squares are either walls, mirrors, or diamonds. One square is the starting position for a ball; another is the target position. We release the ball in one of the four axis-parallel directions, and we may flip mirrors between their two diagonal orientations while the ball moves. The ball reflects at mirrors and stops at walls; at diamonds, it turns around and erases the diamond. The goal is to reach the target position. In this basic form, Reflexion is polynomial; simple variations are NP-complete or PSPACE-complete [100].

Lemmings. *Lemmings* is a computer puzzle in which characters called lemmings start at one or more initial locations and behave according to deterministic rules. The player can modify this basic behavior by applying a skill to a lemming. The goal is for a specified number of lemmings to reach a specified target position; the exact rules are too complicated to detail here. These puzzles are NP-complete, even with just one lemming [29].

A.10   Constraint Logic

For a concise reference to the various constraint-logic problems, see Appendix D.

A.11   Open Problems

In this section we highlight some games whose complexity is unknown or that have some interesting aspect that is open. Many of them were already described in preceding sections. Many of these are likely targets for constraint-logic reductions; some of them have not yielded to attempts to apply constraint logic.

Phutball. *Phutball* was not described in any detail in Section A.6, so we describe it here. Phutball [8] is played on a Go board, with one black stone, the ball, initially placed on the center of the board. On his turn, a player may either place a white stone or make a series of jumps. A jump is made by jumping the ball over a contiguous line of stones, horizontally, vertically, or diagonally, into an empty space. The white stones jumped are removed before the next jump. A player may make as many jumps as he wishes on a single turn. The game is won by jumping the ball onto or over the goal line. Left's goal line is the right edge of the board, and Right's is the left edge of the board.

This game has the unusual property that it is NP-complete merely to determine whether there is a single-move win [38]! The complexity of determining the winner from an arbitrary position was recently shown to be PSPACE-hard [47]. As an unbounded two-player game of perfect information, it could be as hard as EXPTIME-complete. But, as with Lunar Lockout, the nature of the game makes it extremely difficult to construct any sort of stable gadget.

Dots-and-Boxes. *Dots-and-Boxes* (Section A.9) is NP-hard [8, pp. 577–578], but as a two-player bounded game, by all rights it should be PSPACE-complete. Probably it will be easier to show a generalized version, *Strings-and-Coins* (Section A.9), PSPACE-complete, but no reduction method presents itself.

Hackenbush. *Hackenbush* (Section A.9) is NP-hard [8, pp. 189–227], but as a bounded two-player game, it ought to be PSPACE-complete.

Col. *Col* (Section A.3.2) appears not to have been studied from a complexity perspective, though a related game, Snort, is PSPACE-complete [147]. It could potentially yield to a straightforward reduction.

Domineering. *Domineering (Crosscram)* [8] is played on a rectangular grid. Two players take turns placing dominoes on the grid, covering two unoccupied squares; one player can only place dominoes aligned horizontally, and the other only vertically. The first player unable to move loses. This very natural game has a geometric structure that suggests a simple constraint-logic reduction, and we and others have attempted to show it hard. However, there are as yet no complexity results (though there are many interesting positive results—efficient algorithms for special cases—in [8]).

All of the games so far in this section are introduced or discussed in *Winning Ways*, the definitive work (along with *On Numbers and Games*) in combinatorial game theory [8]. *Winning Ways* is filled with many other games, too numerous to list here, that have yet to be studied from a complexity perspective.

Another large source of games waiting to be tackled is Nikoli's pencil puzzles. Many of these have been shown NP-complete (Section A.7). Pencil puzzles for which we know of no complexity results include *Hashiwokakero (Bridges)*, *Kuromasu ("Where is Black Cells")*, *Number Link*, *Ripple Effect*, *Shikaku*, and *Yajilin (Arrow Ring)*. Others are listed on Nikoli's Japanese website.

Lunar Lockout. *Lunar Lockout* (Section A.2.1) is an unbounded puzzle and so is potentially PSPACE-complete. It is known to be NP-hard [97]. The fact that the blocks are 1×1 makes it very difficult to build gadgets, as is also the case with 1×1 Rush Hour, below. However, there is some hope for a reduction from Nondeterministic Constraint Logic to Lunar Lockout; the challenges with 1×1 Rush Hour appear to be more severe. Adding immovable blocks to Lunar Lockout does enable a PSPACE-completeness proof [82].

1×1 Rush Hour. 1×1 *Rush Hour* (Section A.2.1) empirically seems hard: the minimum solution length seems to grow exponentially with the puzzle size [164]. However, there seems to be little hope for a reduction from Nondeterministic Constraint Logic or indeed for any reduction at all where car movements represent propagating signals. The reason is that it is always easy to determine whether a given car can move at all. This problem seems to straddle the line between easy and hard problems, and as a result is very tantalizing.

Subway Shuffle. *Subway Shuffle* (Section A.3.1) is a generalization of 1×1 Rush Hour. The goal is to move a given token to a given vertex in a graph, where the tokens and edges are colored, and a move is to slide a token from one vertex to an adjacent unoccupied vertex, along an edge of the same color as the token. The game 1×1 Rush Hour is Subway Shuffle restricted to grid graphs with two token colors, with all horizontal edges one color, and all vertical edges the other. The extra flexibility in Subway Shuffle puzzles should make it easier to build hardness gadgets. Therefore, it is even more frustrating and tantalizing that no hardness proof has been found. Subway Shuffle suffers from the same problem as 1×1 Rush Hour: it is easy to determine whether a token can move.

Push-1. *Push-1* (Section A.2.2) is NP-hard; is it PSPACE-complete? Intermediate steps to showing this would be to show PSPACE-completeness for Push-1-F or Push-2; Push-2-F is PSPACE-complete (Section 9.8). Other block-pushing puzzles with unresolved complexity include Push-*, Push-Push-*, and Push-1-G (Section A.2.2).

Retrograde Chess. The *Retrograde Chess* problem is as follows: given two generalized chess positions (configurations of chess pieces on an $n \times n$ board, with one king per player), is it possible to play from one configuration to the other if the players cooperate? This problem is known to be NP-hard [13]; is it PSPACE-complete? This problem seems to almost, but not quite, yield to a Nondeterministic Constraint Logic reduction.

Peg Duotaire and HotSpot. *Peg Duotaire* and *HotSpot* (Section A.4) are two natural peg-jumping games that also appear not to have been studied from a complexity perspective.

Dominono. *Dominono* [73, Chapter 26] is a relative of Tic-Tac-Toe, played on a square grid. Two players, X and O, take turns placing their mark in empty squares. The first player to form a domino—two adjacent squares filled with his mark—loses. This game could be PSPACE-complete, but it appears to be unstudied from a complexity perspective.

Clobber. *Clobber* is played on a rectangular grid, initially filled in a checkerboard pattern with black and white stones. Players Black and White take turns; on his turn, a player moves one of his stones onto an adjacent stone of the opposite color, "clobbering" it (removing both stones from the board). The first player unable to move loses. A solitaire version of Clobber is NP-complete [41]; the complexity of the two-player game is open.

Twixt. *Twixt* (Section A.5, [50]) could potentially be PSPACE-complete; a related problem (Uncrossed Knights Paths) is NP-complete [119].

Chinese Checkers. *Chinese Checkers* stands out as a popular board game with unaddressed complexity. As an unbounded two-player game it is potentially EXPTIME-complete.

Go. *Go* (Section A.6) with Japanese rules has been "solved," from a complexity standpoint, for 25 years: it is EXPTIME-complete [143]. It is rather remarkable, therefore, that the complexity of Go with the addition of the superko rule, as used, for example, in China and in the USA, is still unresolved.

Robson, and others who have studied the problem (notably John Tromp), are of the opinion that Go with superko is probably in EXPTIME. However, as an unbounded, no-repeat, two-player game it "ought" to be EXPSPACE-complete (Section 6.3). But this may be a case where there is effectively some special structure in the game that makes it easier. The EXPTIME-hardness reduction builds variable gadgets out of sets of kos (basic repeating patterns). If all dynamic states are to be encoded in kos, then the problem is in fact in EXPTIME, because it is an instance of Undirected Vertex Geography (Section A.3.3), which is polynomial in the input size. The input size is exponential in this case; it is the space of possible board configurations. But it may be possible to build gadgets that are not merely sets of kos. It is very difficult to do so; Go is "almost" in PSPACE, because in normal play moves are not reversible, and it is only through capture that there is the possibility of the repeating patterns necessary for a harder complexity.

Bridge. A hand of *Bridge* is a bounded team game with private information. Therefore, determining the winner could potentially be as hard as NEXP-TIME (Section 7.1). A reduction from Bounded Team Private Constraint Logic would appear to be difficult, for two reasons. First, in Bounded TPCL, the private information resides in the moves selected; the initial state is known. But, in Bridge the private information resides in the initial disposition of the cards. Second, there is no natural geometric structure to exploit in Bridge, as there is in a typical board game. Still, it is conceivable there could be some reduction.

Stratego. *Stratego* is an unbounded two-player game with private information. There is not at present a corresponding kind of constraint logic, however such games are in general 2EXPTIME-complete (complete in doubly exponential time) [136].

Rengo Kriegspiel. *Rengo Kriegspiel* is a kind of team, blindfold Go. Four players sit facing away from each other, two Black, two White. A referee sees the actual game board in the center. The players take turns attempting to make moves. Only the referee sees the attempted moves. The referee announces when a move is illegal, or when the team's own stone is already on the intersection, and when captures are made. (He also removes captured stones from all players' boards.) Players gradually gain knowledge of the position as the game progresses and the board fills up.

This is an unbounded team game with private information and therefore could potentially be undecidable (Section 7.2).[2] It would appear to be extremely difficult to engineer a reduction from Team Private Constraint Logic to Rengo Kriegspiel, showing undecidability, but perhaps it is possible. A necessary first step would probably be to strengthen the undecidability result for Team Private Constraint Logic (Section 7.2) to apply for the $k = 1$ case (allowing a single edge to be reversed per turn). Next, it seems likely that constructing any constraint-logic gadgets in Rengo Kriegspiel

---

[2]Technically, Rengo Kriegspiel as played in the USA should be assumed to use a superko rule by default. This prevents the global pattern from ever repeating, and thus bounds the length of the game, making it decidable. However, we can consider the game without superko.

would be at least as hard as constructing them in Go, so perhaps we should expect to see an EXPSPACE-completeness result for Go with superko before we see a reduction for Rengo Kriegspiel.

Nevertheless, Rengo Kriegspiel stands out as a game that humans actually play that is not obviously decidable; we are not aware of any other such game.

Computational-Complexity Reference

This appendix serves as a refresher on computability and complexity, and as a "cheat sheet" for the complexity classes used in this book. For more thorough references see, for example, [123], [151], [132], or [104].

The fundamental model of computation used in computer science is the Turing machine. We begin with a definition of Turing machines. Note that there are generally irrelevant differences in the precise forms that different authors use for defining Turing machines.

B.1 Basic Definitions

Turing Machines. Informally, a Turing machine is a deterministic computing device that has a finite number of states, and a one-way infinite tape in which each tape cell contains a symbol from a finite alphabet. The machine has a scanning head that is positioned over some tape cell. On a time step, the machine writes a new symbol in the currently scanned cell, moves the head left or right, and enters a new state, all as a function of the current state and the current symbol scanned. If the transition function is ever undefined, the machine halts, and either accepts or rejects the computation, based on whether it halted in the special accepting state.

Formally, a Turing machine is a 6-tuple $(Q, \Gamma, \delta, q_0, b, q_{\text{accept}})$ where

- Q is a finite set of *states*,

- Γ is a finite set of *tape symbols*,

- $\delta : Q \times \Gamma \to Q \times \Gamma \times \{L, R\}$, a partial function, is the *transition function*,

- $q_0 \in Q$ is the *initial state*,

- $b \in \Gamma$ is the *blank symbol*, and

- $q_{\text{accept}} \in Q$ is the *accepting state*.

A *configuration* is a string in $(Q \cup \Gamma)^*$ containing exactly one symbol from Q. A configuration represents the contents of the tape, the current state, and the current head position at a particular time; the symbol to the right of the state symbol in the string is the symbol currently scanned by the head. The rest of the tape is empty (filled with blank symbols); configurations identical except for trailing blanks are considered equivalent.

The *next-state relation* $\vdash$ relates configurations separated by one step of the Turing machine. When the head motion, state change, and symbol change following the previous head position from a to b correspond to the transition specified in δ, then $a \vdash b$. The transitive and reflexive closure of $\vdash$ is $\vdash^*$.

A Turing machine *halts* on input $w \in \Gamma^*$ in configuration x if $q_0 w \vdash^* x$ and $\neg \exists y \; x \vdash y$. It *accepts* input w if it halts in configuration x for some x that contains q_{accept}; it *rejects* w if it halts but does not accept.

A Turing machine M *computes function* f_M if M halts on input w in configuration $q_{\text{accept}} x$, where $x = f_M(w)$, for all w.

Languages. A *(formal) language* is a set of strings over some alphabet. The language $\{w \mid M \text{ accepts } w\}$ that a Turing machine M accepts is denoted $L(M)$. If some Turing machine accepts a language L, then L is *Turing recognizable* (also called *recursively enumerable* (RE)). RE is the class of all Turing-recognizable languages.

A language corresponds to a *decision problem*—given a string w, is w in the language?

Decidability. If a Turing machine M halts for every input, then it *decides* its language $L(M)$, and is called a *decider*. If some Turing machine decides a language L, then L is *decidable* (also called *recursive*); otherwise, L is *undecidable*. R is the class of all decidable languages. Note that a Turing machine M that computes a function f_M must be a decider.

One example of an undecidable language is the formal language corresponding to the decision problem, "Given a Turing machine M and input w, does M halt on input w?" This is called the *halting problem*. A string in the actual language would consist of encodings of M and w according to some rule.

Complexity. A Turing machine *uses time t on input w* if it halts on input w in t steps: $w \vdash c_1 \vdash \ldots \vdash c_t$. The *time complexity* of a Turing machine M that is a decider is a function $t(n) = $ the maximum number of steps M uses on any input of length n.

The *time complexity class* $\text{TIME}(t(n))$ is the class of languages decided by some Turing machine with time complexity in $O(t(n))$.

Space complexity is defined similarly. A Turing machine *uses space s on input w* if it halts on input w using configurations with maximum length s (not counting trailing blanks). The *space complexity* of a Turing machine M that is a decider is a function $f(n) = $ the maximum space M uses on any input of length n.

The *space complexity class* $\text{SPACE}(f(n))$ is the class of languages decided by some Turing machine with space complexity in $O(f(n))$.

We are now ready to define some commonly used complexity classes:

$$\text{P} = \bigcup_k \text{TIME}(n^k),$$

$$\text{PSPACE} = \bigcup_k \text{SPACE}(n^k),$$

$$\text{EXPTIME} = \bigcup_k \text{TIME}(2^{n^k})$$

are the classes of languages decidable in, respectively, polynomial time, polynomial space, and exponential time. Another important class, NP, will have to wait for Section B.2 for definition.

Reducibility. A language L is polynomial-time reducible to language L' if there is a Turing machine M with polynomial time complexity such that $w \in L \iff f_M(w) \in L'$. That is, membership of a string in L may be tested by computing a polynomial-time function of the string and testing the result in L'.

Completeness. A language L is *hard* for a complexity class X (abbreviated X-*hard*) if every language $L' \in X$ is polynomial-time reducible to L. A language L is *complete* for a complexity class X (abbreviated X-*complete*) if $L \in X$ and L is X-hard.

Intuitively, the languages that are X-complete are the "hardest" languages in X to decide. For example, if every language in PSPACE can be reduced to a language L in polynomial time, then L must be at least as hard as any other language in PSPACE to decide, because one can always translate such a problem in to a membership test for L. The notion of polynomial-time reducibility is used, because a function that can be computed in polynomial time is considered a "reasonable" function.[1]

Note that if a language L is X-hard and L is polynomial-time reducible to language L', then L' is also X-hard. This fact is the basis for most hardness proofs.

B.2   Generalizations of Turing Machines

The basic one-tape, deterministic Turing machine, as defined above, can be enhanced in various ways. For example, one could imagine a Turing machine with multiple read-write tapes, instead of just one. Are such machines more powerful than the basic machine? In this case, any multitape Turing machine M has an equivalent single-tape machine M' that accepts the same language, with at most a quadratic slowdown. Relative to the above complexity classes, they are the same.

Nondeterminism.   One kind of enhancement that seems to increase the power is *nondeterminism*. A *nondeterministic* Turing machine is defined similarly to a deterministic one, except that the transition function δ is allowed to be multivalued:

$$\delta : Q \times \Gamma \to 2^{Q \times \Gamma \times \{L,R\}}.$$

That is, a nondeterministic transition function specifies an arbitrary set of possible transitions. The above definition of acceptance still works, but the meaning has changed: a nondeterministic Turing machine accepts input w if there is *any* accepting computation history $q_0 w \vdash^* x$. Thus, a nondeterministic computer is allowed to nondeterministically "guess" the sequence of transitions needed to accept its input. If there is no such sequence, then it rejects.

Whether nondeterminism actually increases the power of Turing machines is a very important unresolved question [150].

Nondeterministic Complexity.   By analogy with the above definitions, we can define time- and space- complexity classes for nondeterministic Turing machines.

---

[1]However, this definition is only appropriate for classes harder than P, because any language in P is polynomial-time reducible to any other language in P. To define P-completeness appropriately, we need the notion of *log-space reducibility*, which we will not define. See, e.g., [151] for details.

The *nondeterministic time complexity class* NTIME($t(n)$) is the class of languages decided by some nondeterministic Turing machine with time complexity in $O(t(n))$. The *nondeterministic space complexity class* NSPACE($f(n)$) is the class of languages decided by some nondeterministic Turing machine with space complexity in $O(f(n))$.

We may now define some additional complexity classes:

$$\mathrm{NP} = \bigcup_k \mathrm{NTIME}(n^k),$$

$$\mathrm{NPSPACE} = \bigcup_k \mathrm{NSPACE}(n^k)$$

are the classes of languages decidable in, respectively, nondeterministic polynomial time and nondeterministic polynomial space.

The relationship between P and NP is unknown. Clearly P $\subseteq$ NP, but is NP strictly larger? That is, are there problems that can be solved efficiently—in polynomial time—using a nondeterministic computer, but that cannot be solved efficiently using a deterministic computer? We cannot actually build nondeterministic computers, so the question may seem academic, but many important problems are known to be NP-complete [74], so if P $\neq$ NP, then there is no efficient deterministic algorithm for solving them.

However, it *is* known that PSPACE = NPSPACE [146]. More generally, NSPACE($f(n)$) $\subseteq$ SPACE($f^2(n)$). Nondeterminism thus does not increase the power of space-bounded computation beyond at most a quadratic savings.

In relation to the concept of games and puzzles, a nondeterministic computation is similar to a puzzle: if the right moves to solve the "computation puzzle" may be found, then the computation nondeterministically accepts. We cannot build an actual nondeterministic computer, but we *can* build and solve puzzles. A perfect puzzle solver is performing a nondeterministic computation.

Alternation. Chandra, Kozen, and Stockmeyer [22] have extended the concept of nondeterminism to that of *alternation*. Essentially, the idea is to add the notion of universal, as well as existential, quantification. A nondeterministic Turing machine accepts existentially: it accepts if there exists an accepting computation history. In an alternating Turing machine, the states are divided into *existential* states and *universal* states. A machine accepts starting from a configuration in an existential state if *any* transition from the transition function leads to acceptance; it accepts starting from a configuration in a universal state if *all* possible transitions lead to acceptance.

Alternating time- and space-complexity classes $\text{ATIME}(t(n))$ and $\text{ASPACE}(f(n))$ are defined as above, and AP and APSPACE are defined analogously to P and PSPACE (or NP and NPSPACE). It turns out that AP = PSPACE, and APSPACE = EXPTIME. Thus, alternating time is as powerful as deterministic space, and alternating space is as powerful as exponential deterministic time.

But what does alternation mean, intuitively? The best way to think of an alternating computation is as a two-player game. One player, the existential one, is trying to win the game (accept the computation) by choosing a winning move (transition); the other player, the universal one, is trying to win (reject the computation) by finding a move (transition) from which the existential player cannot win. And in fact, the concept of alternation, and the results mentioned above, have been very useful in the field of game complexity.

Again, we cannot build an alternating computer, but we can play actual two-player games; a perfect game player is performing an alternating computation.

Multiplayer Alternation. Building on the notion of alternation, Peterson and Reif [133] introduced *multiplayer alternation*. It turns out that simply adding new computational "players," continuing the idea that an extra degree of nondeterminism adds computational power, is not sufficient here. Instead, a multiplayer computation is like a *team* game, with multiple players on a team, and with the additional notion of *private information*. The game analogy is that in some games, not all information is public to all players. (Many card games, for example, have this property.) The concept is added to Turing machines by having multiple read-write tapes, with the transition function from some states not allowed to depend on the contents of some tapes. Multiplayer alternation is explored in Chapters 7 and 8.

Multiplayer alternating machines turn out to be extremely powerful—so powerful, in fact, that MPA-PSPACE, the class of languages decidable in multiplayer alternating polynomial space, is all the decidable languages [133].

This is a remarkable fact. A multiplayer alternating Turing machine can do in a bounded amount of space what a deterministic Turing machine can do with an *infinite* tape. Again, we cannot build actual multiplayer alternating computers. But if we lived in a world containing perfect game players, we could do arbitrary computations with finite physical resources.

B.3 Relationship of Complexity Classes

The containment relationships of the classes mentioned above are as follows:

$$P \subseteq NP \subseteq PSPACE = NPSPACE = AP \subseteq EXPTIME = APSPACE \subsetneq R \subsetneq RE$$

All of the containments are believed to be strict, but beyond the above relations, the only strict containment known among those classes is $P \subsetneq$ EXPTIME. $P \overset{?}{=} NP$ is the most famous unknown relation, but it is not even known whether $P = PSPACE$.

B.4 List of Complexity Classes Used in this Book

The following classes are listed in order of increasing containment; that is, $L \subseteq NL \subseteq NC^3 \ldots$, with the exception that the relationship between NP and coNP is unknown. (However, either $NP = coNP$, or neither contains the other.)

L $= SPACE(\log n)$.

NL $= NSPACE(\log n)$.

NC^3 see, e.g., [151] for definition.

P $= \bigcup_k TIME(n^k) =$ languages decidable in polynomial time.

NP $= \bigcup_k NTIME(n^k) =$ languages decidable in nondeterministic polynomial time.

coNP $= \{L \mid \overline{L} \in NP\} =$ languages whose complements are decidable in nondeterministic polynomial time. ($w \in \overline{L} \iff w \notin L$.)

PSPACE $= \bigcup_k SPACE(n^k) =$ languages decidable in polynomial space.

NPSPACE $= \bigcup_k NSPACE(n^k) =$ languages decidable in nondeterministic polynomial space $= PSPACE$.

AP $= \bigcup_k ATIME(n^k) =$ languages decidable in alternating polynomial time $= PSPACE$.

EXPTIME $= \bigcup_k TIME(2^{n^k}) =$ languages decidable in exponential time.

APSPACE $= \bigcup_k ASPACE(n^k) =$ languages decidable in alternating polynomial space $= EXPTIME$.

NEXPTIME $= \bigcup_k NTIME(2^{n^k}) =$ languages decidable in nondeterministic exponential time.

EXPSPACE $= \bigcup_k \text{SPACE}(2^{n^k}) =$ languages decidable in exponential space.

NEXPSPACE $= \bigcup_k \text{NSPACE}(2^{n^k}) =$ languages decidable in nondeterministic exponential space $=$ EXPSPACE.

2EXPTIME $= \bigcup_k \text{TIME}(2^{2^{n^k}}) =$ languages decidable in doubly exponential time.

R $=$ decidable languages.

RE $=$ Turing-recognizable languages.

B.5   Formula Games

A game played on a Boolean formula is often the canonical complete problem for a complexity class. Boolean Satisfiability (SAT), the first problem shown to be NP-complete [28], can be viewed as a puzzle in which the moves are to choose variable assignments. Quantified Boolean Formulas (QBF), which is PSPACE-complete, essentially turns this puzzle into a two-player game, where the players alternate choosing variable assignments. There are formula games for EXPTIME and EXPSPACE, and other classes, as well.

Here we define Boolean formulas and discuss the basic formula games SAT and QBF. Other formula games are defined in the text as they are needed, and many are listed in Section A.8.

B.5.1   Boolean Formulas

A *Boolean variable* is a variable that can have the value *true* or *false*. A *Boolean operation* is one of AND ($\wedge$), OR ($\vee$), or NOT ($\neg$). A *Boolean formula* is either a Boolean variable, or one of the expressions $(\phi \wedge \psi)$, $(\phi \vee \psi)$, and $\neg\phi$, where ϕ and ψ are Boolean formulas.

Expression $(\phi \wedge \psi)$ is true if ϕ and ψ are both true, and false otherwise; $(\phi \vee \psi)$ is true if either ϕ or ψ is true, and false otherwise; $\neg\phi$ is true if ϕ is false, and false otherwise.

A *literal* is a variable x or its negation $\neg x$, abbreviated $\overline{x}$.

A *monotone formula* is a formula that does not contain $\neg$. Monotone formulas have the property that if the value of any contained variable is changed from false to true, the value of the formula can never change from true to false.

A *(disjunctive) clause* is either a literal or $(\phi \vee \psi)$, where ϕ and ψ are disjunctive clauses; a *conjunctive clause* is either a literal or $(\phi \wedge \psi)$, where ϕ and ψ are conjunctive clauses.

A Boolean formula F is in *conjunctive normal form* (CNF) if it is either a clause or $(\phi \wedge \psi)$, where ϕ and ψ are in CNF; F is in *disjunctive normal form* (DNF) if it is either a conjunctive clause or $(\phi \vee \psi)$, where ϕ and ψ are in DNF. F is in kCNF (kDNF) if it is in CNF (DNF) and each of its clauses contains at most k literals.

A *quantified variable* is either $\forall x$ or $\exists x$, for variable x.

A *quantified Boolean formula* is either a Boolean formula, or a quantified Boolean formula preceded by a quantified variable.

Expression $\forall x \, \phi$ is true if ϕ is true both when x is assigned to false and when it is assigned to true, and $\exists x \, \phi$ is true if ϕ is true when x is assigned either to false or to true.

B.5.2   Satisfiability (SAT)

The Boolean formula satisfiability problem is NP-complete and is almost invariably the problem of choice to which to reduce another problem to show that it is NP-hard. It is defined as follows:

SATISFIABILITY (SAT)

Instance: Boolean formula ϕ.

Question: Is there an assignment to the variables of ϕ such that ϕ is true?

Equivalently, SAT could be defined as the question of whether a given quantified Boolean formula that uses only existential quantifiers is true. The process of choosing a satisfying variable assignment can be viewed as solving a kind of puzzle.

In 3SAT, ϕ is in 3CNF; 3SAT is also NP-complete.

B.5.3   Quantified Boolean Formulas (QBF)

Quantified Boolean Formulas is PSPACE-complete and is almost invariably the problem of choice to which to reduce another problem to show that it is PSPACE-hard. It is defined as follows:

QUANTIFIED BOOLEAN FORMULAS (QBF)

Instance: Quantified Boolean formula ϕ.

Question: Is ϕ true?

The truth of a quantified Boolean formula corresponds to the winner of a two-person game. This is easiest to see in the case where the quantifiers strictly alternate between $\exists$ and $\forall$, as in $\exists x \, \forall y \, \exists z \ldots \phi$. Then, we may say that the $\exists$ player can win the formula game if he can choose a value for x such that for any value the $\forall$ player chooses for y, the $\exists$ player can choose a value for z, such that $\ldots \phi$ is true.

This correspondence may also be understood in terms of the previously mentioned result that AP = PSPACE: a two-player game of polynomially bounded length is an alternating computation that can be carried out in polynomial time.

QBF remains PSPACE-complete if ϕ is in 3CNF.

Deterministic Constraint Logic Activation Sequences

In this appendix, we present the explicit activation sequences for several DCL gadgets described in Section 4.2. Refer to that section for complete descriptions of the gadgets' intended behaviors, and of the deterministic rule used. As mentioned there, all of the gadgets used are designed on the assumption that signals will only arrive their inputs at some time 0 mod 4 (so that the first internal edge reversal occurs at time 1 mod 4), and also that signals will activate output edges only at times 0 mod 4. This makes it possible to know what the internal state of the gadgets is when inputs arrive, because any persistent activity in a gadget will repeat every two or four steps.

Switch Gadget. This gadget is used internally in many of the other gadgets. In Figure C.1, we show all the steps in its activation sequence. When input arrives at A, an output signal is sent first to B, then, when that signal has returned to the switch, on to C, then to B again, and finally back to A. In some cases the extra activation of B is useful; in the other cases, it is redundant but not harmful.

Existential Quantifier. This gadget uses a switch to "try" both possible variable assignments. The connected CNF circuitry follows a protocol by which a variable state can be asserted by activating one pathway; a return pathway will then activate back to the quantifier. When the quantifier is done

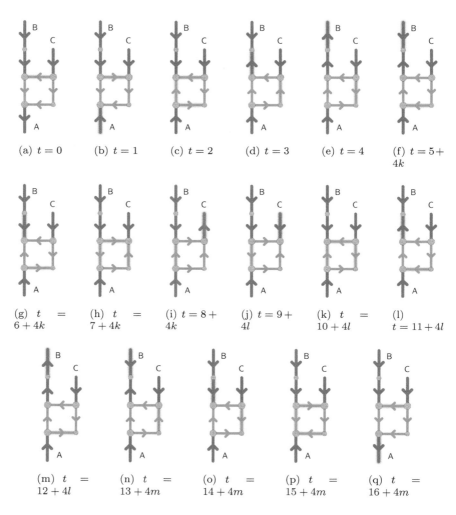

Figure C.1. Switch gadget steps. $0 \leq k \leq l \leq m$.

using that assignment, it can de-assert it by following the return pathway backward; activation will then proceed back into the gadget along the original assertion output edge.

In Figure C.2 we show the activation sequence. Only every fourth time step is shown; in between these steps the internal switch is operating as above. Possible activation of the satisfied in/satisfied out pathway is not shown, but when it occurs it clearly preserves the necessary timing.

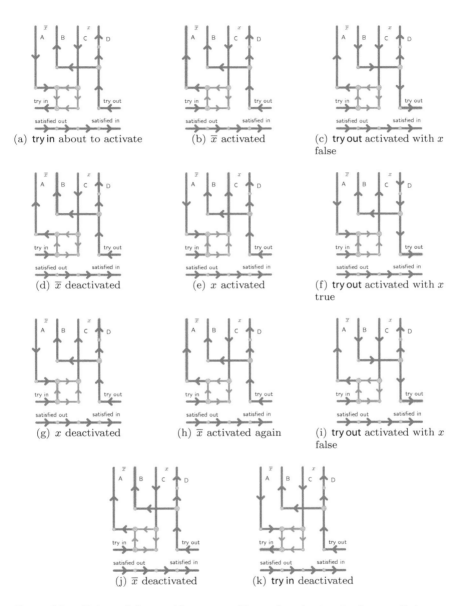

Figure C.2. Existential-quantifier steps. Every fourth step is shown. Between steps (b) and (c), (d) and (e), (e) and (f), (g) and (h), (h) and (i), and (j) and (k), a signal is propagated into and out of the connecting CNF circuitry. Between steps (c) and (d), (f) and (g), and (i) and (j), a signal is propagated through to the quantifier to the right, and possibly through satisfied in/satisfied out and back. All inputs are guaranteed to arrive at times 0 mod 4.

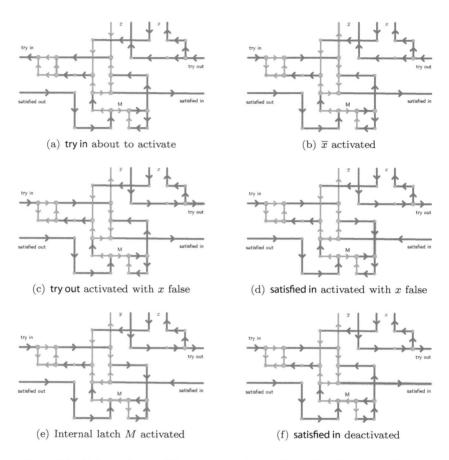

(a) **try in** about to activate   (b) $\bar{x}$ activated

(c) **try out** activated with x false   (d) **satisfied in** activated with x false

(e) Internal latch M activated   (f) **satisfied in** deactivated

Figure C.3. Universal-quantifier steps, part one. Every fourth step is shown.

Universal Quantifier. The universal quantifier is similar to the existential quantifier; it also uses a switch to try both variable assignments. However, if the assignment to x = false succeeds, the gadget sets an internal latch to remember this fact. Then, if x = true also succeeds, the latch enables **satisfied out** to be directed out. Finally, the switch tries x = false again; this resets the latch. This sequence is shown in Figures C.3 and C.4. Again, only every fourth time step is shown. Only the "forward" operation of the gadget is shown; deactivation follows an inverse sequence.

If the assignment to x = false fails, and x = true succeeds, then the unset latch state causes the x = true success to simply bounce back. This sequence is shown in Figure C.5.

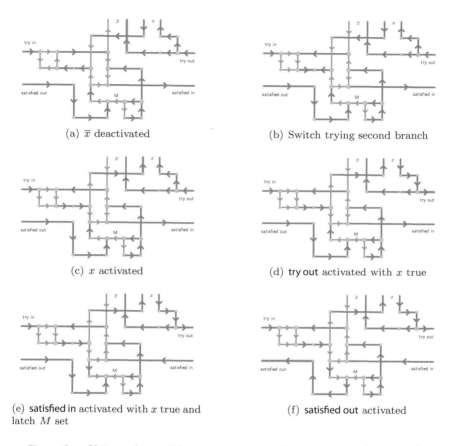

(a) $\bar{x}$ deactivated

(b) Switch trying second branch

(c) x activated

(d) **try out** activated with x true

(e) **satisfied in** activated with x true and latch M set

(f) **satisfied out** activated

Figure C.4. Universal-quantifier steps, part two (continuation of part one).

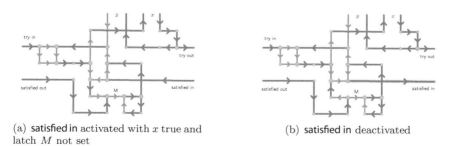

(a) **satisfied in** activated with x true and latch M not set

(b) **satisfied in** deactivated

Figure C.5. Universal-quantifier steps, part three. These two steps replace the last two in Figure C.4, in the case where the x false assignment did not succeed and set latch M.

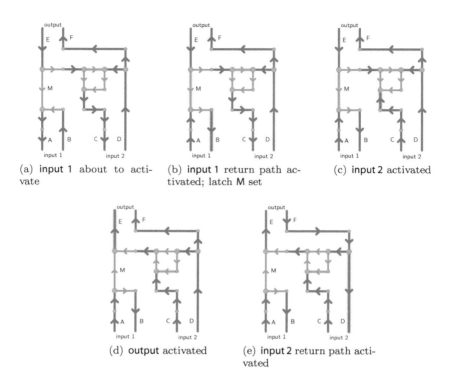

(a) input 1 about to acti-   (b) input 1 return path ac-   (c) input 2 activated
vate   tivated; latch M set

(d) output activated   (e) input 2 return path acti-
                              vated

Figure C.6. AND′ steps, in the case when both inputs activate in sequence. Every fourth step is shown.

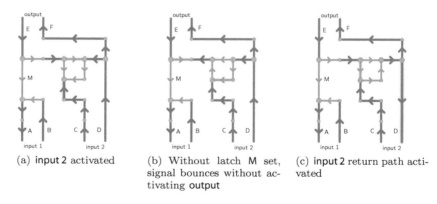

(a) input 2 activated   (b) Without latch M set,   (c) input 2 return path acti-
                         signal bounces without ac-   vated
                         tivating output

Figure C.7. AND′ steps, in the case when input 2 activates without input 1 first activating. Every fourth step is shown.

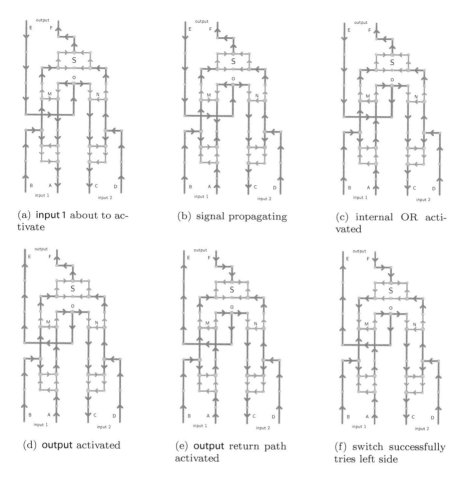

(a) **input 1** about to activate

(b) signal propagating

(c) internal OR activated

(d) **output** activated

(e) **output** return path activated

(f) switch successfully tries left side

Figure C.8.   OR′ steps, part one. Every fourth step is shown. If **input 2** is activated instead, the sequence will be slightly different; the switch will first try the left side, and then the right.

AND′. The AND′ gadget must respond to two different circumstances: first, **input 1** arrives, and then **input 2** later arrives (or not); and second, **input 2** arrives when **input 1** has not arrived. The first case is shown in Figure C.6, the second in Figure C.7. In each case only every fourth time step is shown, and the reverse, deactivating, sequences are not shown.

OR′. The OR′ is complicated for two reasons. First, it must activate when either input activates, but whichever has activated, if the other input then

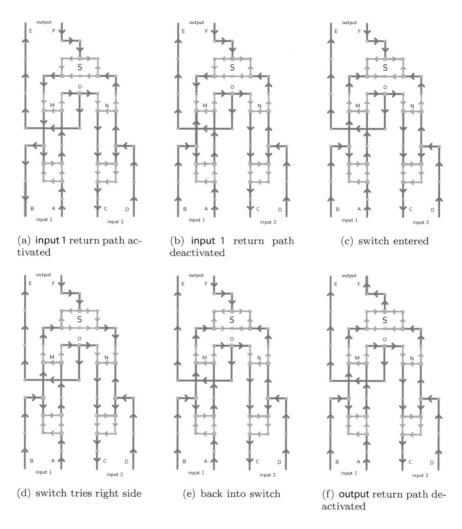

(a) input 1 return path activated

(b) input 1 return path deactivated

(c) switch entered

(d) switch tries right side

(e) back into switch

(f) output return path deactivated

Figure C.9. OR′ steps, part two. Every fourth step is shown. If input 2 is activated instead, the sequence will be slightly different; the switch will send the return signal out earlier.

arrives, it must simply bounce back cleanly (because the output is already activated). Second, the internal switch required is more complicated than the basic switch. The basic switch may be described as following the sequence ABCBA; the switch used in the OR′ would correspondingly follow the sequence ABC.

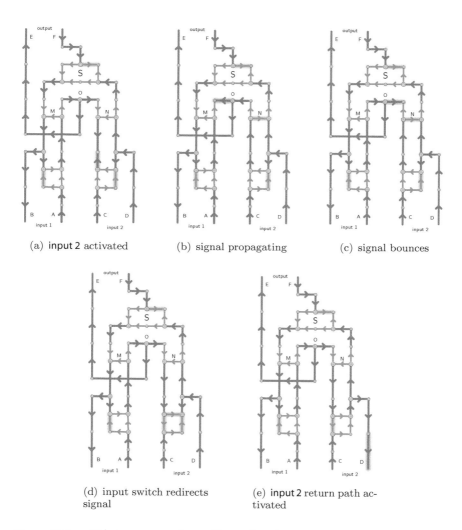

(a) **input 2** activated   (b) signal propagating   (c) signal bounces

(d) input switch redirects signal

(e) **input 2** return path activated

Figure C.10.   OR′ steps, part three. Every fourth step is shown; **input 2** arrives when the gate is already activated, and is cleanly propagated on to its return path. Deactivation follows the reverse sequence.

Two sequences are shown. First, in Figures C.8 and C.9, an activation sequence beginning with **input 1** is shown. Part of the deactivating sequence is shown as well, because it is not the reverse of the forward sequence (due to the modified switch). The activation sequence beginning with **input 2** is similar, but the "extra search step" taken by the internal switch

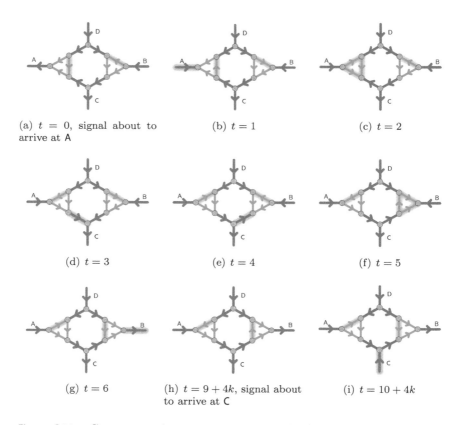

(a) $t = 0$, signal about to arrive at A

(b) $t = 1$

(c) $t = 2$

(d) $t = 3$

(e) $t = 4$

(f) $t = 5$

(g) $t = 6$

(h) $t = 9 + 4k$, signal about to arrive at C

(i) $t = 10 + 4k$

Figure C.11.   Crossover gadget steps, part one. The padding edges required to enter and exit at times 0 mod 4 are omitted; as a result, there is a gap between steps (g) and (h).

occurs during the forward rather than the reverse activation sequence in this case.

Second, Figure C.10 shows the activation sequence when the OR′ is already active, and the other input arrives. (In this case the operation is symmetric with respect to the two inputs.) The second input is propagated directly to its return path.

FANOUT′, CNF Output Gadget. The correct operation of these gadgets (shown in Figure 4.5) is obvious.

Crossover Gadget. The steps involved in crossover activation are shown in Figures C.11 and C.12. The reverse sequence deactivates the crossover.

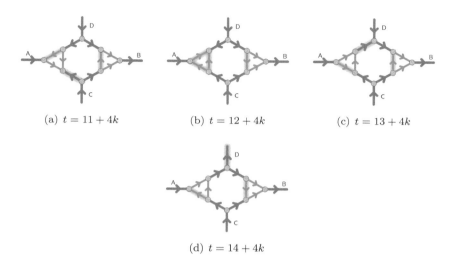

(a) $t = 11 + 4k$ (b) $t = 12 + 4k$ (c) $t = 13 + 4k$

(d) $t = 14 + 4k$

Figure C.12. Crossover gadget steps, part two (continuation of part one).

The sequence shown has a **C-D** traversal following an **A-B** traversal. **C-D** can also occur in isolation (but not followed by **A-B**); note that after the **A-B** traversal (and at the same time mod 4), the gadget is in a vertically symmetric state to the original one.

Constraint-Logic Quick Reference

Table D.1 provides a concise reference to all of the complexity results for different varieties of constraint logic. This table is the starting point when trying to show that a new problem is hard: one can quickly identify the set of gadgets that must be built for a reduction and identify what variants of the appropriate constraint logic are also hard.

In the table, each kind of constraint logic is complete for the indicated class, for planar constraint graphs, except as noted. Generally, there is a simple set of "basis vertex" types that may be used for reductions. The basis vertices listed in the table are a sufficient set to implement such reductions. Not all of these are AND or OR vertices; the normal definitions of constraint logic (Section 2.1) apply. Degree-3 vertices have an inflow constraint of 2; degree-2 vertices have an inflow constraint of 1.

As remarked in Section 2.1 (page 19), for all the different kinds of constraint logic, it is sufficient to consider constraint graphs containing only AND and OR vertices. However, using only ANDs and ORs for game reductions could mean implementing a large number of vertex subtypes. For example, for Bounded Two-Player constraint logic (Bounded 2CL), there are 42 functionally distinct types of AND (red-red-blue) and OR (blue-blue-blue) vertex, depending on the initial orientation and controlling player assigned to each edge. Reducing Bounded 2CL for general AND and OR graphs to a given problem would mean building gadgets for each of these. Instead, it is sufficient to implement the listed basis vertices.

Constraint Logic	Complete for Class	Basis Vertices	Comments
Zero Player, Bounded (Bounded DCL, Section 4.1)	P		Result does not hold for planar graphs.
Zero Player, Unbounded (DCL, Section 4.2)	PSPACE		
One Player, Bounded (Bounded NCL, Section 5.1)	NP		Constraint Graph Satisfiability is also NP-complete (Section 5.1.3).
One Player, Unbounded (NCL, Section 5.2)	PSPACE		Remains PSPACE-complete when all ORs are protected (Section 5.2.3). Configuration-to-configuration problem is also PSPACE-complete (Section 5.2.4).
Two Player, Bounded (Bounded 2CL, Section 6.1)	PSPACE		
Two Player, Unbounded (2CL, Section 6.2)	EXPTIME		
Team, Private, Bounded (Bounded TPCL, Section 7.1)	NEXPTIME	see Section 7.1	
Team, Private, Unbounded (TPCL, Section 7.2)	RE	see Section 7.2	k edges may be reversed in one turn.

Table D.1. Constraint-Logic Quick Reference.

Bibliography

[1] Scott Aaronson. "NP-complete Problems and Physical Reality." *SIGACT News* 36:1 (2005), 30–52.

[2] Hiroyuki Adachi, Hiroyuki Kamekawa, and Shigeki Iwata. "Shogi on $n \times n$ Board Is Complete in Exponential Time (in Japanese)." *Transactions of the IEICE* J70-D:10 (1987), 1843–1852.

[3] Michael H. Albert, Richard J. Nowakowski, and David Wolfe. *Lessons in Play: An Introduction to Combinatorial Game Theory.* Wellesley, MA: A K Peters, Ltd., 2007.

[4] Daniel Andersson. "HIROIMONO Is NP-Complete." In *Fun with Algorithms: 4th International Conference, FUN 2007, Castiglioncello, Italy, June 3–5, 2007, Proceedings,* Lecture Notes in Computer Science 4475, pp. 30–39. Berlin: Springer, 2007.

[5] László Babai. "Trading Group Theory for Randomness." In *Proceedings of the Seventeenth Annual ACM Symposium on Theory of Computing,* pp. 421–429. New York: ACM Press, 1985.

[6] Charles H. Bennett. "Logical Reversibility of Computation." *IBM J. Res. & Dev.* 17 (1973), 525–532.

[7] Robert Berger. "The Undecidability of the Domino Problem." *Memoirs of the American Mathematical Society* 66 (1966), 1–72.

[8] Elwyn R. Berlekamp, John H. Conway, and Richard K. Guy. *Winning Ways for Your Mathematical Plays,* four volumes, Second edition. Wellesley, MA: A K Peters, Ltd., 2001–2004.

[9] Elwyn Berlekamp. *The Dots and Boxes Game: Sophisticated Child's Play.* Wellesley, MA: A K Peters, Ltd., 2000.

[10] Elwyn R. Berlekamp. "Sums of $N \times 2$ Amazons." In *Game Theory, Optimal Stopping, Probability and Statistics*, Lecture Notes – Monograph Series 35, edited by F. Thomas Bruss and Lucien Le Cam, pp. 1–34. Beachwood, OH: Institute of Mathematical Statistics, 2000.

[11] Therese C. Biedl, Erik D. Demaine, Martin L. Demaine, Rudolf Fleischer, Lars Jacobsen, and J. Ian Munro. "The Complexity of Clickomania." In *More Games of No Chance*, edited by R. J. Nowakowski, pp. 389–404. Cambridge, UK: Cambridge University Press, 2002.

[12] Therese Biedl. "The Complexity of Domino Tiling." In *Proceedings of the 17th Canadian Conference on Computational Geometry*, pp. 187–190, 2005. Available at http://www.cccg.ca/proceedings/2005/6.pdf.

[13] Hans Bodlaender. "Re: Is Retrograde Chess NP-hard?" Usenet posting to rec.games.abstract, 2001.

[14] Paul S. Bonsma and Luis Cereceda. "Finding Paths between Graph Colourings: PSPACE-Completeness and Superpolynomial Distances." In *Mathematical Foundations of Computer Science 2007: 32nd International Symposium, MFCS 2007 Ceský Krumlov, Czech Republic, August 26-31, 2007, Proceedings*, Lecture Notes in Computer Science 4708, pp. 738–749. Berlin: Springer, 2007.

[15] David Bremner, Joseph O'Rourke, and Thomas Shermer. "Motion Planning amidst Movable Square Blocks is PSPACE-complete." Manuscript, 1994.

[16] Ron Breukelaar, Erik D. Demaine, Susan Hohenberger, Hendrik Jan Hoogeboom, Walter A. Kosters, and David Liben-Nowell. "Tetris Is Hard, Even to Approximate." *International Journal of Computational Geometry and Applications* 14:1–2 (2004), 41–68.

[17] John Bruno and Louis Weinberg. "A Constructive Graph-Theoretic Solution of the Shannon Switching Game." *IEEE Transactions on Circuit Theory* CT-17 1 (1970), 74–81.

[18] Kevin Buchin and Maike Buchin. "Rolling Block Mazes Are PSPACE-complete." Manuscript, 2007.

[19] Kevin Buchin, Maike Buchin, Erik D. Demaine, Martin L. Demaine, Dania El-Khechen, Sándor Fekete, Christian Knauer, André Schulz, and Perouz Taslakian. "On Rolling Cube Puzzles." In *Proceedings of the 19th Canadian Conference on Computational Geometry*, pp. 141–144, 2007. Available at http://cccg.ca/proceedings/2007/05b5full.pdf.

[20] Michael Buro. "Simple Amazons Endgames and Their Connection to Hamilton Circuits in Cubic Subgrid Graphs." In *Computers and Games: Second International Conference, CG 2000 Hamamatsu, Japan, October 26–28, 2000 Revised Papers*, Lecture Notes in Computer Science 2063, edited by Tony Marsland and Ian Frank, pp. 250–260. Berlin: Springer, 2000.

[21] Alice Chan and Alice Tsai. "$1 \times n$ Konane: A Summary of Results." In *More Games of No Chance*, edited by R. J. Nowakowski, pp. 331–339. Cambridge, UK: Cambridge Universtiy Press, 2002.

[22] Ashok K. Chandra, Dexter Kozen, and Larry J. Stockmeyer. "Alternation." *Journal of the ACM* 28:1 (1981), 114–133.

[23] Xi Chen and Xiaotie Deng. "Settling the Complexity of Two-Player Nash Equilibria." In *Proceedings of the 47th Annual IEEE Symposium on Foundations of Computer Science*, pp. 261–272. Los Alamitos, CA: IEEE Press, 2006.

[24] Anne Condon and Richard E. Ladner. "Probabilistic Game Automata." *Journal of Computer and System Sciences* 36:3 (1988), 452–489.

[25] Anne Condon, Joan Feigenbaum, and Carsten Lund Peter Shor. "Random Debaters and the Hardness of Approximating Stochastic Functions." *SIAM Journal on Computing* 26:2 (1997), 369–400.

[26] Anne Condon. "Computational Models of Games." Ph.D. thesis, University of Washington, 1988. Published by MIT Press in 1989 as an ACM Distinguished Dissertation.

[27] John Horton Conway. *On Numbers and Games*, Second edition. Natick, MA: A K Peters, Ltd., 2000.

[28] Stephen A. Cook. "The Complexity of Theorem-Proving Procedures." In *Proceedings of the Third IEEE Symposium on the Foundations of Computer Science*, pp. 151–158. Los Alamitos, CA: IEEE Computer Society, 1971.

[29] Graham Cormode. "The Hardness of the Lemmings Game, or Oh No, More NP-completeness Proofs." In *Fun with Algorithms: 4th International Conference, FUN 2007, Castiglioncello, Italy, June 3–5, 2007, Proceedings*, Lecture Notes in Computer Science 4475, pp. 65–76. Berlin: Springer, 2004.

[30] Paolo Cotogno. "Hypercomputation and the Physical Church-Turing Thesis." *Brit. J. Philosophy of Science* 54 (2003), 181–223.

[31] Marcel Crâsmaru and John Tromp. "Ladders Are PSPACE-Complete." In *Computers and Games: Second International Conference, CG 2000 Hamamatsu, Japan, October 26–28, 2000, Revised Papers*, Lecture Notes in Computer Science 2063, pp. 241–249. Berlin: Springer, 2000.

[32] Marcel Crâsmaru. "On the Complexity of Tsume-Go." In *Computers and Games: First International Conference, CG'98 Tsukuba, Japan, November 11–12, 1998, Proceedings*, Lecture Notes in Computer Science 1558, pp. 222–231. Berlin: Springer-Verlag, 1999.

[33] J. C. Culberson. "Sokoban is PSPACE-complete." In *Proceedings International Conference on Fun with Algorithms (FUN98)*, pp. 65–76. Waterloo, Ontario, Canada: Carleton Scientific, 1998.

[34] Constantinos Daskalakis, Paul W. Goldberg, and Christos H. Papadimitriou. "The Complexity of Computing a Nash Equilibrium." In *Proceedings of the 38th Annual ACM Symposium on Theory of Computing*, pp. 71–78. New York: ACM Press, 2006.

[35] Erik D. Demaine and Martin L. Demaine. "Jigsaw Puzzles, Edge Matching, and Polyomino Packing: Connections and Complexity." *Graphs and Combinatorics* 23 (Supplement) (2007), 195–208.

[36] Erik D. Demaine and Robert A. Hearn. "Playing Games with Algorithms: Algorithmic Combinatorial Game Theory." In *Games of No Chance 3*, edited by R. J. Nowakowski, 2008. To appear.

[37] Erik D. Demaine and Michael Hoffmann. "Pushing Blocks Is NP-Complete for Noncrossing Solution Paths." In *Proceedings of the 13th Canadian Conference on Computational Geometry*, pp. 65–68. Waterloo, Canada, 2001. Available at http://compgeo.math.uwaterloo.ca/~cccg01/proceedings/long/eddemaine-24711.ps.

[38] Erik D. Demaine, Martin L. Demaine, and David Eppstein. "Phutball Endgames Are NP-hard." In *More Games of No Chance*, edited by R. J. Nowakowski, pp. 351–360. Cambridge, UK: Cambridge University Press, 2002.

[39] Erik D. Demaine, Martin L. Demaine, and Helena A. Verrill. "Coin-Moving Puzzles." In *More Games of No Chance*, edited by R. J. Nowakowski, pp. 405–431. Cambridge, UK: Cambridge University Press, 2002.

[40] Erik D. Demaine, Robert A. Hearn, and Michael Hoffmann. "Push-2-F Is PSPACE-Complete." In *Proceedings of the 14th Canadian Conference on Computational Geometry (CCCG 2002)*, edited by Stephen Wismath, pp. 31–35. Lethbridge, Alberta, Canada, 2002. Available at http://www.dartmouth.edu/~rah/push-2f.pdf.

[41] Erik D. Demaine, Martin L. Demaine, and Rudolf Fleischer. "Solitaire Clobber." *Theoretical Computer Science* 313:3 (2004), 325–338.

[42] Erik D. Demaine, Michael Hoffmann, and Markus Holzer. "PushPush-k Is PSPACE-Complete." In *Proceedings of the 3rd International Conference on FUN with Algorithms*, pp. 159–170. Pisa, Italy: Università di Pisa, 2004.

[43] Erik D. Demaine, Martin L. Demaine, Arthur Langerman, and Stefan Langerman. "Morpion Solitaire." *Theory of Computing Systems* 39:3 (2006), 439–453.

[44] Erik D. Demaine, Martin L. Demaine, Rudolf Fleischer, Robert A. Hearn, and Timo von Oertzen. "The Complexity of Dyson Telescopes." In *Games of No Chance 3*, edited by R. J. Nowakowski, 2008. To appear.

[45] Erik D. Demaine, Robert A. Hearn, and Erez Lieberman. "Evolving Populations and Their Computational Ability." Manuscript in preparation, 2008.

[46] Dariusz Dereniowski. "The Complexity of Node Blocking for Dags." Manuscript, 2008. Available at http://arxiv.org/abs/0802.3513.

[47] Dariusz Dereniowski. "Phutball Is PSPACE-hard." arXiv:0804.1777, 2008. Available at http://arXiv.org/abs/0804.1777.

[48] Dorit Dor and Uri Zwick. "SOKOBAN and Other Motion Pplanning Problems." *Computational Geometry: Theory and Applications* 13:4 (1999), 215–228.

[49] Henry Ernest Dudeney. *The Canterbury Puzzles and Other Curious Problems*. London: W. Heinemann, 1907.

[50] David Eppstein. "Computational Complexity of Games and Puzzles." Available at http://www.ics.uci.edu/~eppstein/cgt/hard.html, 2007.

[51] David Eppstein. "On the NP-Completeness of Cryptarithms." *SIGACT News* 18:3 (1987), 38–40.

[52] Michael D. Ernst. "Playing Konane Mathematically: A Combinatorial Game-Theoretic Analysis." *UMAP Journal* 16:2 (1995), 95–121.

[53] Shimon Even and Robert E. Tarjan. "A Combinatorial Problem which Is Complete in Polynomial Space." *Journal of the ACM* 23:4 (1976), 710–719.

[54] Hugh Everett. "Relative State Formulation of Quantum Mechanics." *Rev. Mod. Phys.* 29 (1957), 454–462.

[55] U. Feige and A. Shamir. "Multi-oracle Interactive Protocols with Space Bounded Verifiers." In *Proceedings of the Fourth Annual Structure in Complexity Theory Conference*, pp. 158–164. Los Alamitos, CA: IEEE Press, 1989.

[56] Gary William Flake and Eric B. Baum. "*Rush Hour* is PSPACE-complete, or 'Why You Should Generously Tip Parking Lot Attendants'." *Theoretical Computer Science* 270:1–2 (2002), 895–911.

[57] Aviezri S. Fraenkel and Elisheva Goldschmidt. "PSPACE-Hardness of Some Combinatorial Games." *Journal of Combinatorial Theory*, Series A 46 (1987), 21–38.

[58] Aviezri S. Fraenkel and David Lichtenstein. "Computing a Perfect Strategy for $n \times n$ Chess Requires Time Exponential in n." *Journal of Combinatorial Theory*, Series A 31 (1981), 199–214.

[59] A. S. Fraenkel and Y. Yesha. "Complexity of Problems in Games, Graphs and Algebraic Equations." *Discrete Applied Mathematics* 1 (1979), 15–30.

[60] Aviezri S. Fraenkel, Edward R. Scheinerman, and Daniel Ullman. "Undirected Edge Geography." *Theoretical Computer Science* 112:2 (1993), 371–381.

[61] Michael P. Frank. "Approaching the Physical Limits of Computing." In *Proceedings of the 35th IEEE International Symposium on Multiple-Valued Logic (ISMVL 2005)*, pp. 168–185. Los Alamitos, CA: IEEE Computer Society, 2005.

[62] Greg N. Frederickson. *Hinged Dissections: Swinging and Twisting.* Cambridge, UK: Cambridge University Press, 2002.

[63] E. Fredkin and T. Toffoli. "Conservative Logic." *International Journal of Theoretical Physics* 21 (1982), 219–253.

[64] Erich Friedman. "Cubic Is NP-complete." Paper presented at the 34th Annual Florida MAA Section Meeting, Ft. Myers, FL, March 2–3, 2001.

[65] Erich Friedman. "Corral Puzzles Are NP-complete." Unpublished manuscript, 2002. Available at http://www.stetson.edu/~efriedma/papers/corral/corral.html.

[66] Erich Friedman. "Pearl Puzzles Are NP-complete." Unpublished manuscript, 2002. Available at http://www.stetson.edu/~efriedma/papers/pearl/pearl.html.

[67] Erich Friedman. "Pushing Blocks in Gravity Is NP-hard." Unpublished manuscript, 2002. Available at http://www.stetson.edu/~efriedma/papers/gravity/gravity.html.

[68] Erich Friedman. "Spiral Galaxies Puzzles Are NP-complete." Unpublished manuscript, 2002. Available at http://www.stetson.edu/~efriedma/papers/spiral/spiral.html.

[69] Timothy Furtak, Masashi Kiyomi, Takeaki Uno, and Michael Buro. "Generalized Amazons Is PSPACE-Complete." In *Proceedings of the Nineteenth International Joint Conference on Artificial Intelligence*, pp. 132–137. Denver, CO: Professional Book Center, 2005.

[70] Zvi Galil. "Hierarchies of Complete Problems." *Acta Informatica* 6:1 (1976), 77–88.

[71] Martin Gardner. "The Hypnotic Fascination of Sliding-Block Puzzles." *Scientific American* 210 (1964), 122–130. Also Chapter 7 of *Martin Gardner's Sixth Book of Mathematical Diversions*, University of Chicago Press, Chicago, 1984.

[72] Martin Gardner. "Mathematical Games: The Fantastic Combinations of John Conway's New Solitaire Game 'Life'." *Scientific American* 223:4 (1970), 120–123.

[73] Martin Gardner. *A Gardner's Workout: Training the Mind and Entertaining the Spirit.* Natick, MA: A K Peters, Ltd., 2001.

[74] Michael R. Garey and David S. Johnson. *Computers and Intractability: A Guide to the Theory of NP-Completeness.* New York: W. H. Freeman & Co., 1979.

[75] Andrea Gilbert. "Plank Puzzles." Available at http://www.clickmazes.com/planks/ixplanks.htm, 2000.

[76] Andrea Gilbert. "Wriggle Puzzles." Available at http://www.clickmazes.com/tjwrig/ixjwrig.htm, 2007.

[77] L. M. Goldschlager. "The Monotone and Planar Circuit Value Problems Are log Space Complete for P." *SIGACT News* 9:2 (1977), 25–29.

[78] Arthur S. Goldstein and Edward M. Reingold. "The Complexity of Pursuit on a Graph." *Theoretical Computer Science* 143 (1995), 93–112.

[79] Shafi Goldwasser, Silvio Micali, and Charles Rackoff. "The Knowledge Complexity of Interactive Proof Systems." *SIAM Journal on Computing* 18:1 (1989), 186–208.

[80] Markus Götz. "Triagonal Slide-Out." Available at http://www.markus-goetz.de/puzzle/java/0021/crh.html, 2005.

[81] John Gribbin. "Doomsday Device." *Analog Science Fiction/Science Fact* 105:2 (1985), 120–125.

[82] Jeffrey R. Hartline and Ran Libeskind-Hadas. "The Computational Complexity of Motion Planning." *SIAM Review* 45 (2003), 543–557.

[83] Juris Hartmanis and Richard E. Stearns. "On the Computational Complexity of Algorithms." *Transactions of the American Mathematics Society* 117 (1965), 285–306.

[84] Robert A. Hearn and Erik D. Demaine. "The Nondeterministic Constraint Logic Model of Computation: Reductions and Applications." In *Automata, Languages and Programming: 29th International Colloquium, ICALP 2002, Malaga, Spain, July 8–13, 2002, Proceedings*, Lecture Notes in Computer Science 2380, edited by Peter Widmayer, Francisco Triguero Ruiz, Rafael Morales Bueno, Matthew Hennessy, Stephan Eidenbenz, and Ricardo Conejo, pp. 401–413. Berlin: Springer, 2002.

[85] Robert A. Hearn and Erik D. Demaine. "PSPACE-Completeness of Sliding-Block Puzzles and Other Problems through the Nondeterministic Constraint Logic Model of Computation." *Theoretical Computer Science*, "Game Theory Meets Theoretical Computer Science" Special Issue, 343:1–2 (2005), 72–96.

[86] Robert Hearn, Erik Demaine, and Greg Frederickson. "Hinged Dissection of Polygons Is Hard." In *Procedings of the 15th Canadian Conference on Computational Geometry*, pp. 98–102, 2003. Available at http://www.cccg.ca/proceedings/2003/45.pdf.

[87] Robert A. Hearn. "The Complexity of Sliding Block Puzzles and Plank Puzzles." In *Tribute to a Mathemagician*, edited by Barry Cipra, Erik Demaine, Martin Demaine, and Tom Rodgers, pp. 173–183. Wellesley, MA: A K Peters, Ltd., 2004.

[88] Robert A. Hearn. "Amazons Is PSPACE-complete." Manuscript, 2005. Available at http://www.arXiv.org/abs/cs.CC/0008025.

[89] Robert A. Hearn. "The Subway Shuffle Puzzle." Manuscript, 2005. Available at http://www.subwayshuffle.com.

[90] Robert Hearn. "TipOver Is NP-Complete." *Mathematical Intelligencer* 28:3 (2006), 10–14.

[91] Robert A. Hearn. "Games, Puzzles, and Computation." Ph.D. thesis, Department of Electrical Engineering and Computer Science, Massachusetts Institute of Technology, Cambridge, MA, 2006. Available at http://www.swiss.ai.mit.edu/~bob/hearn-thesis-final.pdf.

[92] Robert A. Hearn. "Amazons, Konane, and Cross Purposes Are PSPACE-complete." In *Games of No Chance 3*, edited by R. J. Nowakowski, 2008. To appear.

[93] Robert A. Hearn. "The Complexity of Minesweeper Revisited." Manuscript in preparation, 2008.

[94] Robert A. Hearn. "Hitori Is NP-complete." Manuscript in preparation, 2008.

[95] Malte Helmert. "Complexity Results for Standard Benchmark Domains in Planning." *Artificial Intelligence* 143:2 (2003), 219–262.

[96] Malte Helmert. "New Complexity Results for Classical Planning Benchmarks." In *Proceedings of the Sixteenth International Conference on Automated Planning and Scheduling (ICAPS 2006)*, pp. 52–61. Menlo Park, CA: AAAI Press, 2006.

[97] Martin Hock. "Exploring the Complexity of the UFO Puzzle." Undergraduate thesis, Carnegie Mellon University, 2001. Available at http://www.cs.cmu.edu/afs/cs/user/mjs/ftp/thesis-02/hock.ps.

[98] Michael Hoffmann. "Push-* Is NP-hard." In *Proceedings of the 12th Canadian Conference on Computational Geometry*, pp. 205–210, 2000. Available at http://www.cs.unb.ca/conf/cccg/eProceedings/13.ps.gz.

[99] Markus Holzer and Oliver Ruepp. "The Troubles of Interior Design—A Complexity Analysis of the Game Heyawake." In *Fun with Algorithms: 4th International Conference, FUN 2007, Castiglioncello, Italy, June 3–5, 2007, Proceedings*, Lecture Notes in Computer Science 4475, pp. 198–212. Berlin: Springer, 2007.

[100] Markus Holzer and Stefan Schwoon. "Assembling Molecules in ATOMIX Is Hard." *Theoretical Computer Science* 313:3 (2004), 447–462.

[101] Markus Holzer and Stefan Schwoon. "Reflections on REFLEXION—Computational Complexity Considerations on a Puzzle Game." In *Proceedings of the 3rd International Conference on FUN with Algorithms*, pp. 90–105. Pisa, Italy: Università di Pisa, 2004.

[102] Markus Holzer, Andreas Klein, and Martin Kutrib. "On the NP-completeness of the NURIKABE Pencil Puzzle and Variants Thereof." In *Proceedings of the 3rd International Conference on FUN with Algorithms*, pp. 77–89. Pisa, Italy: Università di Pisa, 2004.

[103] J. E. Hopcroft, J. T. Schwartz, and M. Sharir. "On the Complexity of Motion Planning for Multiple Independent Objects: PSPACE-Hardness of the 'Warehouseman's Problem'." *International Journal of Robotics Research* 3:4 (1984), 76–88.

[104] John E. Hopcroft, Rajeev Motwani, and Jeffrey D. Ullman. *Introduction to Automata Theory, Languages, and Computation*, Second edition. Reading, MA: Addison Wesley, 2000.

[105] Clay Mathematics Institute. "Millennium Problems." Available at http://www.claymath.org/millennium/, 2000.

[106] Shigeki Iwata and Takumi Kasai. "The Othello Game on an $n \times n$ Board Is PSPACE-complete." *Theoretical Computer Science* 123 (1994), 329–340.

[107] Richard Kaye. "Minesweeper Is NP-Complete." *Mathematical Intelligencer* 22:2 (2000), 9–15.

[108] David Kempe. "On the Complexity of the Reflections Game." Unpublished manuscript, 2003. Available at http://www-rcf.usc.edu/~dkempe/publications/reflections.pdf.

[109] F. G. König, M. E. Lübbecke, R. H. Möhring, G. Schäfer, and I. Spenke. "Solutions to Real-World Instances of PSPACE-Complete Stacking." Technical Report 2007/4, Institut für Mathematik, Technische Universtät Berlin, 2007. Available at http://www.math.tu-berlin.de/~luebbeck/papers/stacking.pdf.

[110] Daniel Král, Vladan Majerech, Jiří Sgall, Tomáš Tichý, and Gerhard Woeginger. "It Is Tough to be a Plumber." *Theoretical Computer Science* 313:3 (2004), 473–484.

[111] Richard E. Ladner and Jeffrey K. Norman. "Solitaire Automata." *Journal of Computer and System Sciences* 30:1 (1985), 116–129.

[112] R. Landauer. "Irreversibility and Heat Generation in the Computing Process." *IBM J. Res. & Dev.* 5 (1961), 183.

[113] K. Li and K. H. Cheng. "Complexity of Resource Allocation and Job Scheduling Problems on Partitionable Mesh Connected Systems." In *Proceedings of the 1st Annual IEEE Symposium on Parallel and Distributed Processing*, pp. 358–365. Los Alamitos, CA: IEEE Computer Society, 1989.

[114] David Lichtenstein and Michael Sipser. "GO Is Polynomial-Space Hard." *Journal of the Association for Computing Machinery* 27:2 (1980), 393–401.

[115] David Lichtenstein. "Planar Formulae and Their Uses." *SIAM J. Comput.* 11:2 (1982), 329–343.

[116] Jens Lieberum. "An Evaluation Function for the Game of Amazons." *Theoretical Computer Science*, "Advances in Computer Games" Special Issue, 349:2 (2005), 230–244.

[117] Luc Longpré and Pierre McKenzie. "The Complexity of Solitaire." In *Mathematical Foundations of Computer Science 2007: 32nd International Symposium, MFCS 2007 Ceský Krumlov, Czech Republic, August 26–31, 2007, Proceedings*, Lecture Notes in Computer Science 4708, pp. 182–193. Berlin: Springer, 2007.

[118] Oriel Maxime. "Wriggler Puzzles Are PSPACE-complete." Manuscript, 2007.

[119] D. Mazzoni and K. Watkins. "Uncrossed Knight Paths Is NP-complete." Manuscript, 2007. Available at http://www.math.uni-bielefeld.de/~sillke/PROBLEMS/Twixt_Proof_Draft.

[120] Brandon McPhail. "The Complexity of Puzzles." Undergraduate thesis, Reed College, Portland, Oregon, 2003. Available at http://www.cs.umass.edu/~mcphailb/papers/2003thesis.pdf.

[121] Brandon McPhail. "Light Up Is NP-Complete." Unpublished manuscript, 2005. Availavle at http://www.cs.umass.edu/~mcphailb/papers/2005lightup.pdf.

[122] Brandon McPhail. "Metapuzzles: Reducing SAT to Your Favorite Puzzle." CS Theory talk, 2007. Available at http://www.cs.umass.edu/~mcphailb/papers/2007metapuzzles.pdf.

[123] Marvin Minsky. *Computation: Finite and Infinite Machines.* Englewoods Cliffs, NJ: Prentice Hall, 1967.

[124] Cristopher Moore and David Eppstein. "One-Dimensional Peg Solitaire, and Duotaire." In *More Games of No Chance*, edited by R. J. Nowakowski, pp. 341–350. Cambridge, UK: Cambridge University Press, 2002.

[125] Cristopher Moore and John Michael Robson. "Hard Tiling Problems with Simple Tiles." *Discrete and Computational Geometry* 26:4 (2001), 573–590.

[126] Martin Müller and Theodore Tegos. "Experiments in Computer Amazons." In *More Games of No Chance*, edited by R. J. Nowakowski, pp. 243–257. Cambridge, UK: Cambridge University Press, 2002.

[127] John Nash. "Non-cooperative Games." *Annals of Mathematics, Second Series* 54 (1951), 286–295.

[128] Richard J. Nowakowski, editor. *Games of No Chance.* Cambridge, UK: Cambridge University Press, 1996.

[129] Richard J. Nowakowski, editor. *More Games of No Chance.* Cambridge, UK: Cambridge University Press, 2002.

[130] Richard J. Nowakowski, editor. *Games of No Chance 3.* To appear, 2008.

[131] Christos H. Papadimitriou. "Games Against Nature." *Journal of Computer and System Sciences* 31:2 (1985), 288–301.

[132] Christos H. Papadimitriou. *Computational Complexity.* Reading, MA: Addison Wesley, 1993.

[133] Gary L. Peterson and John H. Reif. "Multiple-Person Alternation." In *Proceedings of the 20th Annual Symposium on Foundations of Computer Science*, pp. 348–363. Los Alamitos, CA: IEEE Press, 1979.

[134] Gary Peterson, John Reif, and Salman Azhar. "Lower Bounds for Multi-player Non-cooperative Games of Incomplete Information." *Computers and Mathematics with Applications* 41 (2001), 957–992.

[135] Daniel Ratner and Manfred Warmuth. "The $(n^2 - 1)$-Puzzle and Related Relocation Problems." *Journal of Symbolic Computation* 10 (1990), 111–137.

[136] John H. Reif. "Universal Games of Incomplete Information." In *Proceedings of the Eleventh Annual ACM Symposium on Theory of Computing*, pp. 288–308. New York: ACM Press, 1979.

[137] John H. Reif. "The Complexity of Two-Player Games of Incomplete Information." *Journal of Computer and System Sciences* 29:2 (1984), 274–301.

[138] John Reif. Personal communication, 2006.

[139] Stefan Reisch. "Gobang ist PSPACE-vollständig." *Acta Informatica* 13 (1980), 59–66.

[140] Stefan Reisch. "Hex ist PSPACE-vollständig." *Acta Informatica* 15 (1981), 167–191.

[141] Paul Rendell. "Turing Universality of the Game of Life." In *Collision-Based Computing*, edited by Andrew Adamatzky, pp. 513–539. London: Springer-Verlag, 2002.

[142] Edward Robertson and Ian Munro. "NP-completeness, Puzzles and Games." *Utilitas Mathematica* 13 (1978), 99–116.

[143] J. M. Robson. "The Complexity of Go." In *Proceedings of the IFIP 9th World Computer Congress on Information Processing*, pp. 413–417. Amsterdam: North-Holland, 1983.

[144] J. M. Robson. "Combinatorial Games with Exponential Space Complete Decision Problems." In *Mathematical Foundations of Computer Science 1984: 11th Symposium, Praha, Czechoslovakia, September 3–7, 1984, Proceedings*, Lecture Notes in Computer Science 176, pp. 498–506. London: Springer-Verlag, 1984.

[145] J. M. Robson. "*N* by *N* Checkers Is EXPTIME Complete." *SIAM Journal on Computing* 13:2 (1984), 252–267.

[146] Walter J. Savitch. "Relationships between Nondeterministic and Deterministic Tape Complexities." *Journal of Computer and System Sciences* 4:2 (1970), 177–192.

[147] Thomas J. Schaefer. "On the Complexity of Some Two-Person Perfect-Information Games." *Journal of Computer and System Sciences* 16 (1978), 185–225.

[148] Merlijn Sevenster. "Battleships as a Decision Problem." *ICGA Journal* 27:3 (2004), 142–147.

[149] Peter W. Shor. "Polynomial-Time Algorithms for Prime Factorization and Discrete Logarithms on a Quantum Computer." *SIAM Journal on Computing* 26:5 (1997), 1484–1509.

[150] Michael Sipser. "The History and Status of the P versus NP Question." In *Proceedings of the Twenty-Fourth Annual ACM Symposium on Theory of Computing*, pp. 603–618. New York: ACM Press, 1992.

[151] Michael Sipser. *Introduction to the Theory of Computation*, Second edition. Florence, KY: Course Technology, 2005.

[152] Jerry Slocum and Dic Sonneveld. *The 15 Puzzle*. Beverly Hills, CA: Slocum Puzzle Foundation, 2006.

[153] Raymond Georg Snatzke. "New Results of Exhaustive Search in the Game Amazons." *Theoretical Computer Science* 313:3 (2004), 499–509.

[154] Paul Spirakis and Chee Yap. "On the Combinatorial Complexity of Motion Coordination." Report 76, Computer Science Department, New York University, 1983.

[155] Michael John Spriggs. "Morphing Parallel Graph Drawings." PhD dissertation, University of Waterloo, 2007. Available at http://uwspace.uwaterloo.ca/handle/10012/3083.

[156] James W. Stephens. "The Kung Fu Packing Crate Maze." Available at
http://www.puzzlebeast.com/crate/index.html, 2003.

[157] Larry J. Stockmeyer and Ashok K. Chandra. "Provably Difficult Combinatorial Games." *SIAM Journal on Computing* 8:2 (1979), 151–174.

[158] L. J. Stockmeyer and A. R. Meyer. "Word Problems Requiring Exponential Time: Preliminary Report." In *Proceedings of the 5th Annual ACM
Symposium on Theory of Computing*, pp. 1–9. New York: ACM Press, 1973.

[159] L. J. Stockmeyer. "The Polynomial-Time Hierarchy." *Theoretical Computer
Science* 3:1 (1976), 1–22.

[160] W. E. Story. "Note on the '15' Puzzle." *American Mathematical Monthly*
2 (1879), 399–404.

[161] Jeff Stuckman and Guo qiang Zhang. "Mastermind Is NP-Complete." *IN-
FOCOMP J. Comput. Sci* 5 (2006), 25–28.

[162] Max Tegmark and Nick Bostrom. "Is a Doomsday Catastrophe Likely?"
Nature 438 (2005), 754.

[163] TOMY. "Eternity II." Available at http://uk.eternityii.com/, 2007.

[164] John Tromp and Rudy Cilibrasi. "Limits of Rush Hour Logic Complexity." arXiv:cs.CC/0502068, 2005. Available at http://arXiv.org/abs/cs.CC/
0502068.

[165] John Tromp. "On Size 2 Rush Hour Logic." Manuscript, 2000. Available
at http://turing.wins.uva.nl/~peter/teaching/tromprh.ps.

[166] A. M. Turing. "On Computable Numbers, with an Application to the
Entscheidungsproblem." *Proceedings of the London Mathematical Society 2*
42 (1937), 230–65. Correction in volume 43.

[167] Nobuhisa Ueda and Tadaaki Nagao. "NP-completeness Results for NONO-
GRAM via Parsimonious Reductions." Technical Report TR96-0008, Department of Computer Science, Tokyo Institute of Technology, Tokyo, Japan,
1996.

[168] Ryuhei Uehara and Shigeki Iwata. "Generalized Hi-Q Is NP-complete."
Transactions of the IEICE E73 (1990), 270–273.

[169] John von Neumann and Oskar Morgenstern. *Theory of Games and Economic Behavior.* Princeton, NJ: Princeton University Press, 1944.

[170] Robert T. Wainwright. "Life Is Universal!" In *Proceedings of the 7th
Conference on Winter Simulation*, 2, 2, pp. 449–459. New York: ACM Press,
1974.

[171] Gordon Wilfong. "Motion Planning in the Presence of Movable Obstacles."
Annals of Mathematics and Artificial Intelligence 3:1 (1991), 131–150.

[172] David Wolfe. "Go Endgames Are PSPACE-hard." In *More Games of
No Chance*, edited by R. J. Nowakowski, pp. 125–136. Cambridge, UK:
Cambridge University Press, 2002.

[173] Stephen Wolfram. *Cellular Automata and Complexity: Collected Papers.*
New York: Perseus Press, 1994.

[174] Honghua Yang. "An NC Algorithm for the General Planar Monotone Circuit Value Problem." In *Proceedings of the Third IEEE Symposium on Parallel and Distributed Processing*, pp. 196–203. Los Alamitos, CA: IEEE Press, 1991.

[175] Takayuki Yato and Takahiro Seta. "Complexity and Completeness of Finding Another Solution and Its Application to Puzzles." *IEICE Transactions on Fundamentals of Electronics, Communications, and Computer Sciences* E86-A:5 (2003), 1052–1060.

[176] Takayuki Yato. "Complexity and Completeness of Finding Another Solution and Its Application to Puzzles." Master's thesis, University of Tokyo, Tokyo, Japan, 2003.

[177] Masaya Yokota, Tatsuie Tsukiji, Tomohiro Kitagawa, Gembu Morohashi, and Shigeki Iwata. "Exptime-completeness of Generalized Tsume-Shogi (in Japanese)." *Transactions of the IEICE* J84-D-I:3 (2001), 239–246.

Index

In this index, boldface denotes a definition.

Lightning Source UK Ltd.
Milton Keynes UK
UKHW020046200223
417168UK00001B/5